LIVING BUILDING CHALLENGE℠ 3.0

A Visionary Path to a Regenerative Future

Printed in Canada

NOTIFICATION

For information, address:
The International Living Future Institute
1501 East Madison Street, Suite 150
Seattle, WA 98122

IT IS TIME TO IMAGINE A LIVING FUTURE AND A WORLD OF LIVING BUILDINGS

IMAGINE a building designed and constructed to function as elegantly and efficiently as a flower: a building informed by its bioregion's characteristics, that generates all of its own energy with renewable resources, captures and treats all of its water, and that operates efficiently and for maximum beauty.

IMAGINE a city block or a college campus sharing resources from building to building, growing food, and functioning without a dependency on fossil fuel-based transportation.

IMAGINE true sustainability in our homes, workplaces, neighborhoods, villages, towns and cities—Socially Just, Culturally Rich and Ecologically Restorative℠.

Ethnobotanical Garden at Bertschi School, Seattle, WA
Living Certification - Living Building Challenge 2.0
Photo: GGLO

The International Living Future Institute issues a challenge:

TO ALL DESIGN PROFESSIONALS, CONTRACTORS AND BUILDING OWNERS to create the foundation for a sustainable future in the fabric of our communities.

TO POLITICIANS AND GOVERNMENT OFFICIALS to remove barriers to systemic change, and to realign incentives and market signals that truly protect the health, safety and welfare of people and all beings.

TO ALL OF HUMANITY to reconcile the built environment with the natural environment, into a civilization that creates greater biodiversity, resilience and opportunities for life with each adaptation and development.

INSTEAD OF A WORLD THAT IS MERELY A LESS BAD VERSION OF THE ONE WE CURRENTLY HAVE—WE ASK A SIMPLE AND PROFOUND QUESTION—WHAT DOES GOOD LOOK LIKE?

SETTING THE IDEAL AS THE INDICATOR OF SUCCESS

THE LIVING BUILDING CHALLENGE IS A PHILOSOPHY, CERTIFICATION AND ADVOCACY TOOL FOR PROJECTS TO MOVE BEYOND MERELY BEING LESS BAD AND TO BECOME TRULY REGENERATIVE.

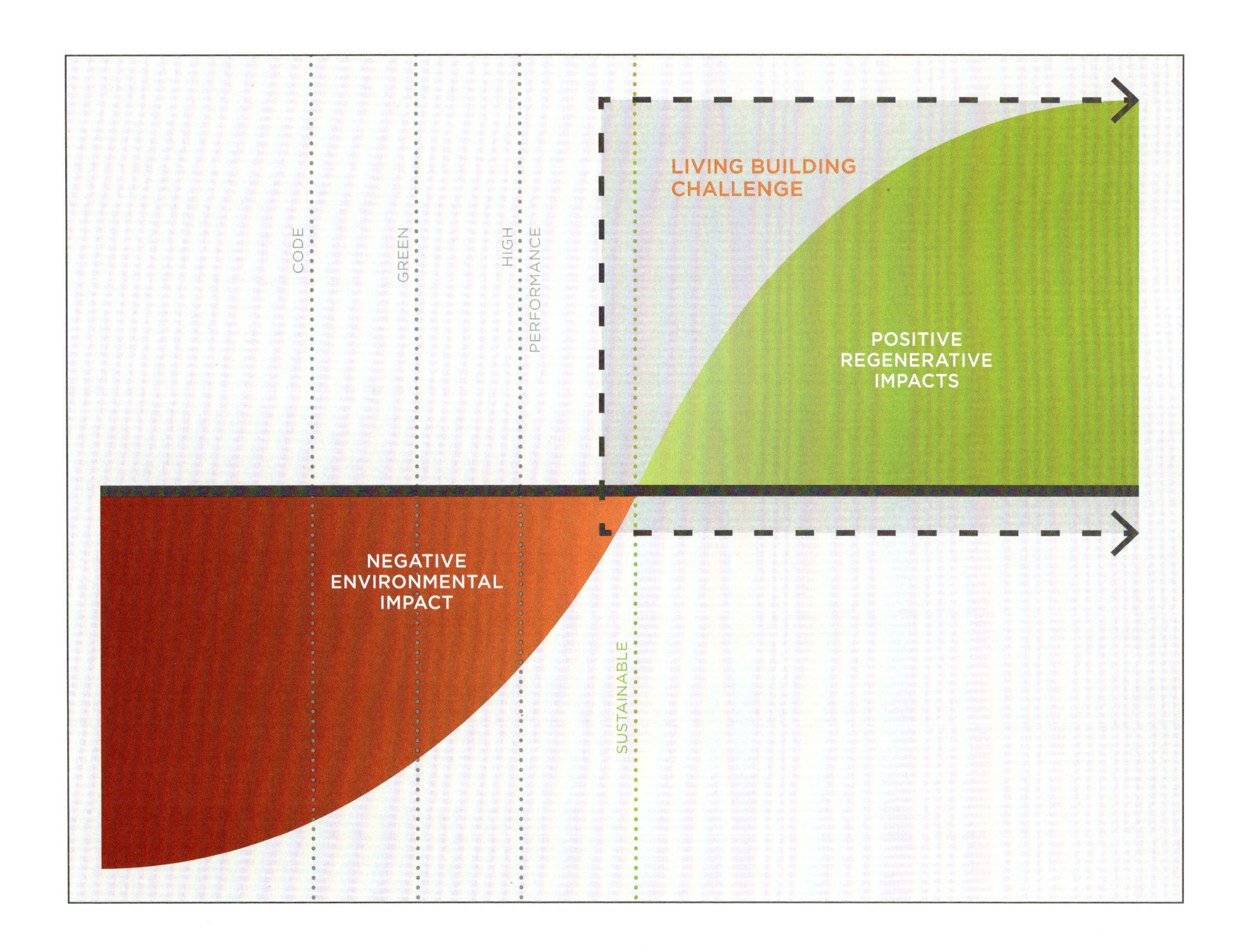

TABLE OF CONTENTS

The Hawaii Preparatory Academy Energy Lab, Kamuela, HI
Living Certification - Living Building Challenge 1.3
Photo: Matthew Millman Photography / Courtesy: Flansburgh Architects

EXECUTIVE SUMMARY: CREATING A REGENERATIVE WORLD TOGETHER

OUR GOAL IS SIMPLE. IN THE WORDS OF BUCKMINSTER FULLER—TO MAKE THE WORLD WORK FOR 100% OF HUMANITY IN THE SHORTEST POSSIBLE TIME THROUGH SPONTANEOUS COOPERATION WITHOUT ECOLOGICAL OFFENSE OR THE DISADVANTAGE OF ANYONE.[1]

The Living Building Challenge is an attempt to dramatically raise the bar from a paradigm of doing less harm to one in which we view our role as steward and co-creator of a true Living Future. The Challenge defines the most advanced measure of sustainability in the built environment possible today and acts to rapidly diminish the gap between current limits and the end-game positive solutions we seek.

The Challenge aims to transform how we think about every single act of design and construction as an opportunity to positively impact the greater community of life and the cultural fabric of our human communities. The program has always been a bit of a Trojan horse—a philosophical worldview cloaked within the frame of a certification program. The Challenge is successful because it satisfies our left brain craving for order and thresholds and our right brain intuition that the focus needs to be on our relationship and understanding of the whole of life.

As such the program is a philosophy first, an advocacy tool second and a certification program third. Within the larger Living Future Challenge framework that covers the creation of all human artifacts and edifices, the Living Building Challenge focuses on humanity's largest creations—its buildings. It is in essence a unified tool for transformative thought, allowing us to envision a future that is Socially Just, Culturally Rich and Ecologically Restorative.

Regardless of the size or location of the project, the Living Building Challenge provides a framework for design, construction and the symbiotic relationship between people and all aspects of community. Indeed, "Living Building Challenge" is not a merely a noun that defines the character of a particular solution for development, but is more relevant if classified as a series of verbs—calls for action that describe not only the building of all of humanity's longest-lasting artifacts, but also of the relationships and broader sense of community and connectivity they engender. It is a challenge to immerse ourselves in such a pursuit—and many refer to the ability to do so as a paradigm shift.

1 The Living Building Challenge was the 2012 winner of the Buckminster Fuller Prize, the world's top award for socially responsible design.

continued >>

Projects that achieve Living Building Status can claim to be the greenest anywhere, and will serve as role models for others that follow. Whether the project is restorative, regenerative or operates with a net zero impact, it has a home in the construct of the Living Building Challenge.

Although it may seem to be ambitious to simultaneously achieve all of the requirements of the Living Building Challenge, understanding the Standard and documenting compliance is inherently easy: there are never more than twenty simple and profound Imperatives that must be met for any type of project, at any scale, in any location around the world.

This Standard is decidedly not a checklist of best practices—the Imperatives of the Living Building Challenge are performance-based and position the ideal outcome as an indicator of success.

The specific methodology used to meet the expectations of the Living Building Challenge is not up to our Institute—but rather to the genius of the design teams, owners and occupants themselves, who are expected to make informed and vested decisions appropriate to the project, place and bioregion.

The Living Building Challenge is a holistic standard, pulling together the most progressive thinking from the worlds of architecture, engineering, planning, interiors, landscape design and policy. It challenges us to ask the question:

What if every single act of design and construction made the world a better place? What if every intervention resulted in greater biodiversity; increased soil health; additional outlets for beauty and personal expression; a deeper understanding of climate, culture and place; a realignment of our food and transportation systems; and a more profound sense of what it means to be a citizen of a planet where resources and opportunities are provided fairly and equitably?

A tall order to be sure.

The scale of change we seek is immense. But without recording these utmost visions and clarity of purpose, we as a society will never experience the type of future that is possible and necessary for our long-term survival. It is our belief that only a few decades remain to completely reshape humanity's relationship with nature and realign our ecological footprint to be within the planet's carrying capacity.

Incremental change is no longer a viable option.

Over the last twenty years, green building has grown to become the most important and progressive trend in the building industry. There have been huge steps forward in the design, construction and operation of buildings, and yet when compared with the rate of change that is required to avoid the worst effects of climate change and other global environmental challenges, our progress has been minute and barely recordable.

The Bullitt Center, Seattle, WA
Photo: Benjamin Benschneider

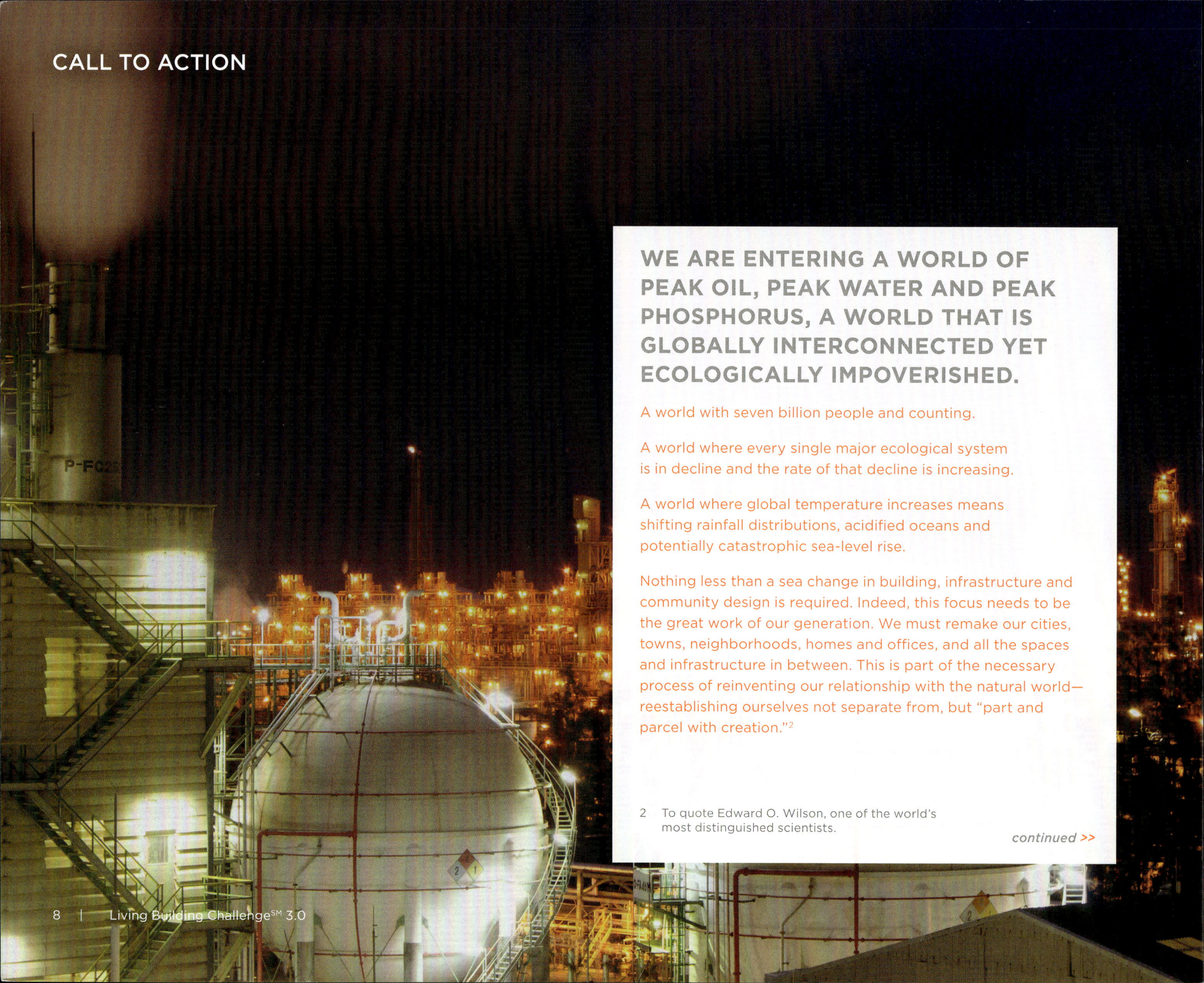

WE ARE ENTERING A WORLD OF PEAK OIL, PEAK WATER AND PEAK PHOSPHORUS, A WORLD THAT IS GLOBALLY INTERCONNECTED YET ECOLOGICALLY IMPOVERISHED.

A world with seven billion people and counting.

A world where every single major ecological system is in decline and the rate of that decline is increasing.

A world where global temperature increases means shifting rainfall distributions, acidified oceans and potentially catastrophic sea-level rise.

Nothing less than a sea change in building, infrastructure and community design is required. Indeed, this focus needs to be the great work of our generation. We must remake our cities, towns, neighborhoods, homes and offices, and all the spaces and infrastructure in between. This is part of the necessary process of reinventing our relationship with the natural world—reestablishing ourselves not separate from, but "part and parcel with creation."[2]

2 To quote Edward O. Wilson, one of the world's most distinguished scientists.

continued >>

Since it was launched in 2006, the Living Building Challenge has inspired and motivated rapid and significant change: projects have sprouted up all over North America and beyond—currently, there are efforts underway in a dozen countries with several million square feet of Living Building Challenge projects in progress—each as beacons in the dark showing what is possible; the regulatory environment has embraced a series of reforms; and most importantly, a new sense of possibility has permeated design communities as a result of the successful certification of the first Living Buildings℠.

THIS STANDARD IS AN ACT OF OPTIMISM AND BELIEF THAT WITH THE RIGHT TOOLS IN THE HANDS OF PASSIONATE, LITERATE AND SENSITIVE INDIVIDUALS, A REVOLUTIONARY TRANSFORMATION IS POSSIBLE. IT IS A PROGRAM THAT ASKS US TO THINK HOLISTICALLY AND ENGAGE BOTH OUR RIGHT AND LEFT BRAINS, HEAD AND HEART.

We invite you to join us, so that together we can continue to forge ahead on our path towards restoration and a Living Future.

PROVEN PERFORMANCE RATHER THAN ANTICIPATED OUTCOMES

The Living Building Challenge is comprised of seven performance categories, or 'Petals': Place, Water, Energy, Health & Happiness, Materials, Equity and Beauty. Petals are subdivided into a total of twenty Imperatives, each of which focuses on a specific sphere of influence. This compilation of Imperatives can be applied to almost every conceivable building project, of any scale and any location—be it a new building or an existing structure.

THERE ARE TWO RULES TO BECOMING A LIVING BUILDING:

- All Imperatives are mandatory. Many of the Imperatives have temporary exceptions to acknowledge current market limitations. These are listed in the Petal Handbooks, which should be consulted for the most up-to-date rulings. Temporary exceptions will be modified or removed as the market changes. With this Standard, the Institute requires advocacy for essential improvements to the building industry.
- Living Building Challenge certification is based on actual, rather than modeled or anticipated, performance. Therefore, projects must be operational for at least twelve consecutive months prior to evaluation for the majority of our Imperative verifications. Some Imperatives can be verified after construction, through a preliminary audit.

The Hawaii Preparatory Academy Energy Lab, Kamuela, HI
Living Certification - Living Building Challenge 1.3
Photo: Matthew Millman Photography / Courtesy: Flansburgh Architects

LIVING CERTIFICATION

A project achieves Living Certification or Living Building Certification by attaining all Imperatives assigned to its Typology. All twenty Imperatives are required for Buildings, fifteen for Renovations and seventeen for Landscape and Infrastructure projects.

PETAL CERTIFICATION

While achieving Living Certification is the ultimate goal, meeting the Imperatives of multiple Petals is a significant achievement in and of itself. Petal Certification requires the achievement of at least three of the seven Petals, one of which must be either the Water, Energy or Materials Petal.

Imperative 01, Limits to Growth and Imperative 20, Inspiration and Education are required.

NET ZERO ENERGY CERTIFICATION

The marketplace has characterized net zero energy in many different ways. The Institute has a simple definition:

One hundred percent of the building's energy needs on a net annual basis must be supplied by on-site renewable energy. No combustion is allowed.

The Net Zero Energy Building Certification program uses the structure of the Living Building Challenge 3.0 to document compliance, it requires four of the Imperatives to be achieved: 01, Limits to Growth, 06, Net Positive Energy (reduced to one hundred percent), 19, Beauty + Spirit, and 20, Inspiration + Education.

The requirement for Imperative 06, Net Positive Energy is reduced to one hundred percent, one hundred and five percent is required for Petal and Living Building Certification only.

As with Living Building and Petal Certification, NZEB certification is based on actual performance rather than modeled outcomes.

David and Lucile Packard Foundation, Los Altos, CA
Net Zero Energy Building Certification
Photo: Terry Lorrant

PETAL CERTIFICATION AND NET ZERO ENERGY BUILDING

David and Lucile Packard Foundation
Los Altos, CA
Photo: Terry Lorrant

zHome, Issaquah, WA
Petal Certification
Photo: zHome

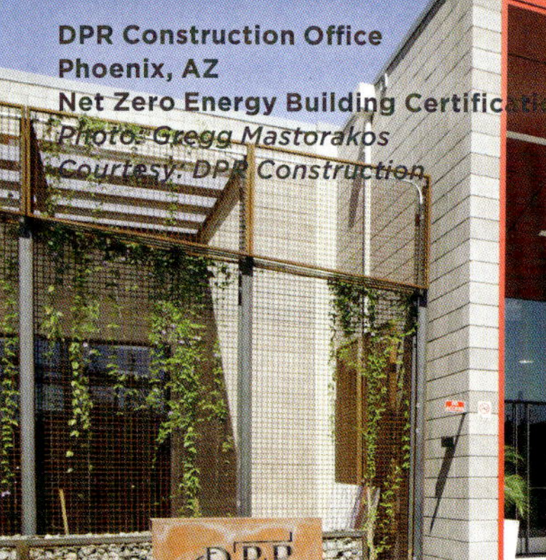

DPR Construction Office
Phoenix, AZ
Net Zero Energy Building Certification
Photo: Gregg Mastorakos
Courtesy: DPR Construction

NRDC Midwest Office
Chicago, IL
Petal Certification
Photo: Studio Gang Architects

CIRS at University of British Columbia
Vancouver, BC
Courtesy: Perkins+Will

June Key Delta Community Center
Portland, OR
Photo: ILFI/Jay Kosa

Living Learning Center at Tyson Research Center, Eureka, MO
Photo: Joe Angeles

VanDusen Botanical Garden, Vancouver, BC
Photo: Nic Lehoux / Courtesy: Perkins+Will

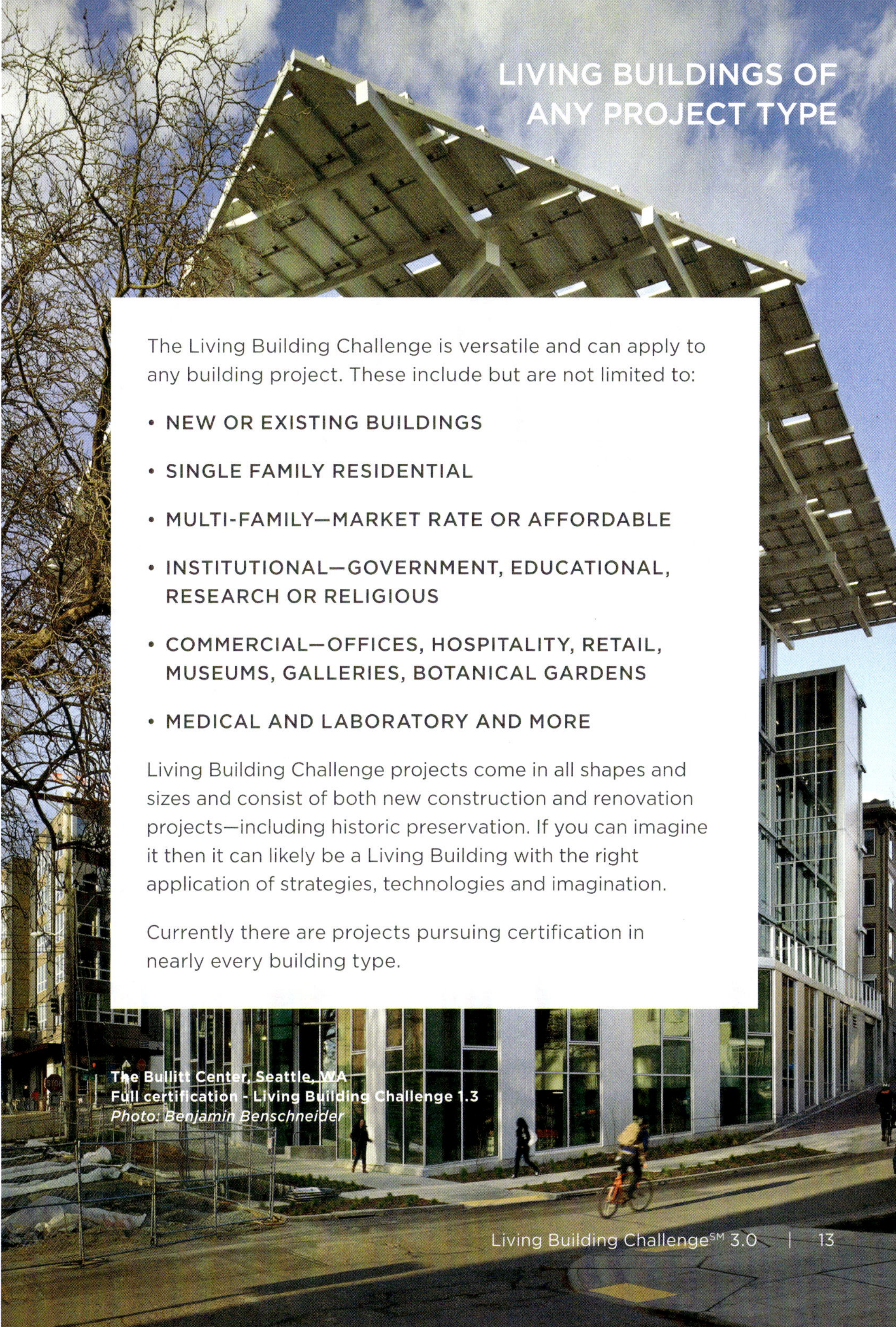

LIVING BUILDINGS OF ANY PROJECT TYPE

The Living Building Challenge is versatile and can apply to any building project. These include but are not limited to:

- NEW OR EXISTING BUILDINGS
- SINGLE FAMILY RESIDENTIAL
- MULTI-FAMILY—MARKET RATE OR AFFORDABLE
- INSTITUTIONAL—GOVERNMENT, EDUCATIONAL, RESEARCH OR RELIGIOUS
- COMMERCIAL—OFFICES, HOSPITALITY, RETAIL, MUSEUMS, GALLERIES, BOTANICAL GARDENS
- MEDICAL AND LABORATORY AND MORE

Living Building Challenge projects come in all shapes and sizes and consist of both new construction and renovation projects—including historic preservation. If you can imagine it then it can likely be a Living Building with the right application of strategies, technologies and imagination.

Currently there are projects pursuing certification in nearly every building type.

The Bullitt Center, Seattle, WA
Full certification - Living Building Challenge 1.3
Photo: Benjamin Benschneider

LIVING BUILDINGS IN EVERY CLIMATE ZONE AND COUNTRY

Living Building Challenge Projects can be built in any climate zone anywhere in the world—as evidenced by the unique array of projects currently underway in many countries around the globe.

This map shows a snapshot of project locations as of March 2014.

Since the Challenge is performance based, the guiding principles and performance metrics apply well regardless of where in the world the project is located—what changes is the specific mix of strategies and technologies—leaving it up to the genius of the design team to choose the most appropriate design response.

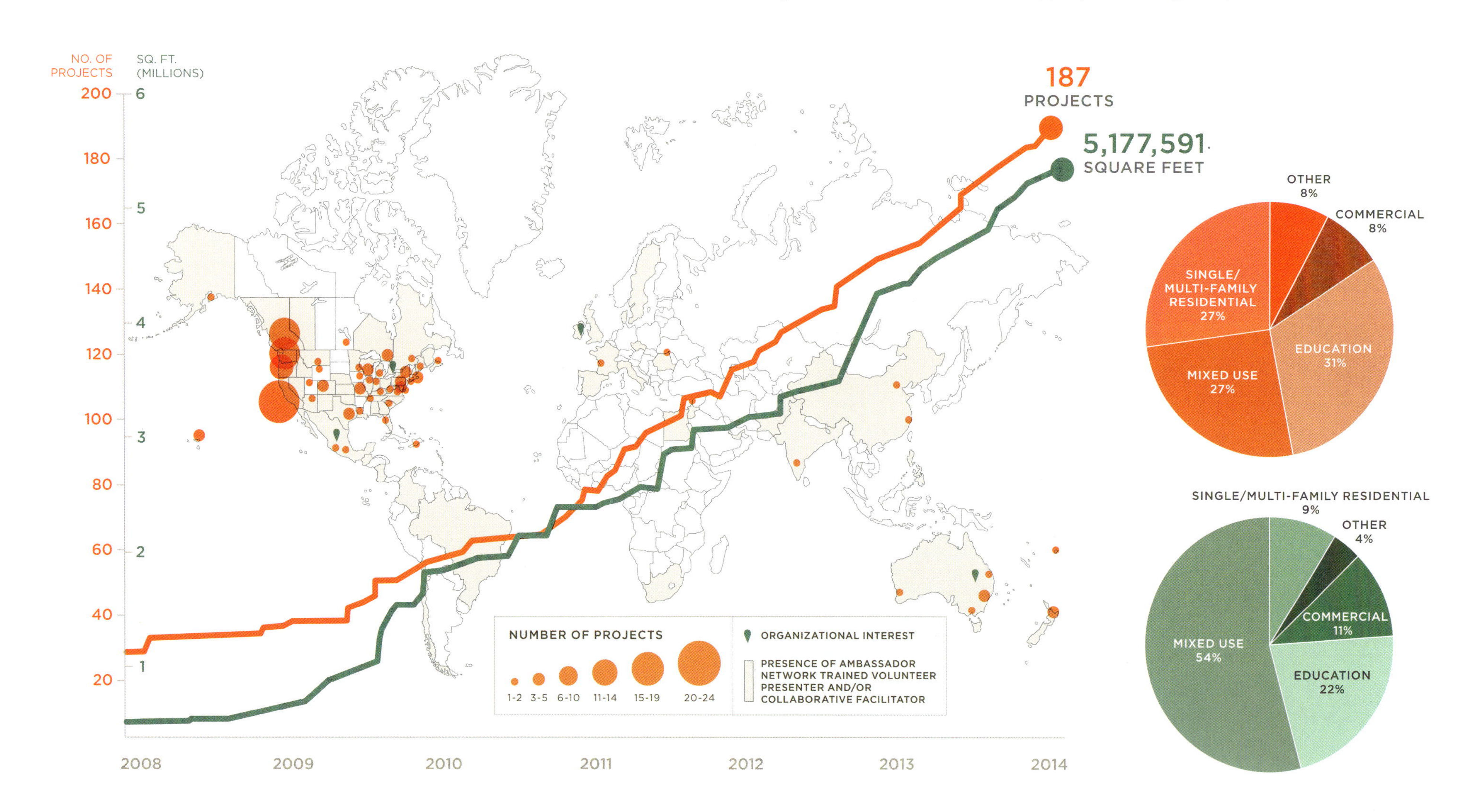

To encourage proper development in specific settings, the standard draws on the work of Duany Plater-Zyberk & Company,[3] who created the New Urbanism Transect model for rural to urban categorization. The Transect is a powerful basis for planning, and demonstrates that different types of standards befit different development realities.[4]

The Living Transect™, which applies to several Imperatives throughout the Living Building Challenge, is an adaptation of the original Transect concept; the significant modification herein is a reclassification of Transect zones T3 and T4 to emphasize appropriate mixed-use densification.

THE CHALLENGE PROMOTES THE TRANSITION OF SUBURBAN ZONES EITHER TO GROW INTO NEW URBAN AREAS WITH GREATER DENSITY, OR TO BE BE DISMANTLED AND REPURPOSED AS NEW RURAL ZONES FOR FOOD PRODUCTION, HABITAT AND ECOSYSTEM SERVICES.

3 www.transect.org
4 These are general descriptions. Refer to the Site Petal Handbook for more information.

continued >>

Photo: joyfull / Shutterstock.com

Photo: Andrew F. Kazmierski / Shutterstock.com

L1. NATURAL HABITAT PRESERVE (GREENFIELD SITES): This transect is comprised of land that is set aside as a nature preserve or is defined as sensitive ecological habitat. It may not be developed except in limited circumstances related to the preservation or interpretation of the landscape, as outlined in the Site Petal Handbook.

L2. RURAL AGRICULTURE ZONE: This transect is comprised of land with a primary function for agriculture and development that relates specifically to the production of food as described in Imperative 02, Urban Agriculture. Small towns and villages do not apply. (Floor Area Ratio ≥ 0.09)

L3. VILLAGE OR CAMPUS ZONE: This transect is comprised of relatively low-density mixed-use development found in rural villages and towns, and may also include college or university campuses. (FAR of 0.1–0.49)

L4. GENERAL URBAN ZONE: This transect is comprised of light- to medium-density mixed-use development found in larger villages, small towns or at the edge of larger cities. (FAR of 0.5–1.49)

L5. URBAN CENTER ZONE: This transect is comprised of a medium- to high-density mixed-use development found in small to mid-sized cities or in the first ‘ring’ of a larger city. (FAR of 1.5–2.99)

L6. URBAN CORE ZONE: This transect is comprised of high-to very high-density mixed use development found in large cities and metropolises. (FAR. ≥ 3.0)

LIVING BUILDING CHALLENGE PROJECTS HAVE THEIR OWN 'UTILITY,' GENERATING THEIR OWN ENERGY AND PROCESSING THEIR OWN WASTE. THEY MORE APPROPRIATELY MATCH SCALE TO TECHNOLOGY AND END USE, AND RESULT IN GREATER SELF-SUFFICIENCY AND SECURITY. YET, THE IDEAL SCALE FOR SOLUTIONS IS NOT ALWAYS WITHIN A PROJECT'S PROPERTY BOUNDARY.

Depending on the technology, the optimal scale can vary when considering environmental impact, first cost and operating costs. To address these realities, the Living Building Challenge has a Scale Jumping overlay to allow multiple buildings or projects to operate in a cooperative state—sharing green infrastructure as appropriate and allowing for Renovation or Building status to be achieved as elegantly and efficiently as possible. Refer to the summary matrix on page 21 to view all Imperatives that may employ the Scale Jumping overlay.[5]

Please note that some projects may then scale from the Living Building Challenge program to the Living Community Challenge program, which are designed to work together.

Imperatives where scale jumping are allowed are marked with this icon.

5 Refer to the handbooks for more information on Scale Jumping.

Omega Institute, Rhinebeck, NY
Living Certification - Living Building Challenge 1.3
Photo: Google Earth

SOME USEFUL GUIDING INFORMATION

- The internal logic of the Living Building Challenge is based on pragmatic, tested experience with what has already been built in the marketplace. Each new Living Building adds further weight to the evidence that a world of Living Buildings is possible now.

- This Standard is an evolving document. Periodically, new releases that update or provide clarification of the Imperatives will be published. Because this Standard is continuously informed by the work that project teams are doing on the ground, Petal Handbooks have been developed to clarify and consolidate the rules at a set point in time to provide a unified reference for project teams. The online Dialogue (see page 65) provides a platform for project teams to request clarifications. A glossary of critical program definitions is provided on page 70.

- The Living Building Challenge does not dwell on basic best-practice issues, so it can instead focus on a smaller number of high level needs. It is assumed that to achieve this progressive standard, typical best practices are being met and championed by the team's expert consultants. The implementation of this standard requires leading-edge technical knowledge, an integrated design approach, and design and construction teams well versed in advanced practices related to green building.

- Regional solutions are manifested in all Living Building Challenge projects due to a number of variables, including climate factors and building characteristics. For example, becoming water-independent in the desert demands evolving a project's design to emulate a cactus instead of a tree. The built environment will be richer because of this response to place.

VanDusen Botanical Garden, Vancouver, BC
Photo: Nic Lehoux / Courtesy: Perkins+Will

THE LIVING BUILDING CHALLENGE IS AN EVER-EVOLVING LIVING PROGRAM SHAPED BY THE INCREDIBLE EXPERIENCES OF OUR PROJECT TEAMS BREAKING NEW GROUND. OVER TIME, FEEDBACK FROM A DIVERSE ARRAY OF STAKEHOLDERS ACTIVELY USING THE CHALLENGE HELPS US UNDERSTAND HOW TO REFINE AND IMPROVE THE PROGRAM. OUR INSTITUTE STAFF ARE ALSO MONITORING CHANGES IN THE FIELD AND THE MARKETS AND MAKING ADJUSTMENTS AS NEEDED TO REFLECT CURRENT REALITIES AND CHALLENGES. THE GOAL IS ALSO TO KEEP RAISING THE BAR AS WE LEARN TOGETHER, MOVING OUR PROJECTS CLOSER STILL TO THE GOAL OF A REGENERATIVE LIVING FUTURE.

continued >>

LIVING BUILDING CHALLENGE 3.0 REPRESENTS AN IMPORTANT STEP FORWARD IN OUR PROGRAM'S EVOLUTION, WITH SEVERAL NEW INNOVATIVE ELEMENTS AS WELL AS IMPORTANT REFINEMENTS.

A FEW KEY CHANGES INCLUDE THE FOLLOWING:

- We have renamed the Site Petal as the Place Petal—reflecting our deeply held belief in viewing each project location as a place with unique and important characteristics, as opposed to merely a site ready for development. Every place is unique and deserves special focus and attention.
- We have removed the community typology from the LBC as it has a new home in the newly launched Living Community Challenge, because we have found that working at scales beyond individual buildings (as well as the infrastructure and urban fabric between buildings) deserves its own focus and attention, and market-based exceptions vary dramatically. Projects may still 'scale jump' to the community scale by using both systems.
- There is a much greater emphasis on the importance of resilient infrastructure—helping to ensure that in a time of uncertainty and disruption, Living Buildings are always beacons of safety and security.
- There is a more explicit emphasis on the idea of the Challenge as a tool for regenerative design, rather than earlier framings of 'do no harm,' which, although always part of our philosophy and goals for the program, was not made clear enough. The LBC is not a net neutral program, it most decidedly is about creating a pathway and vision for a truly sustainable, regenerative living future.
- The LBC has more clearly fleshed out and refined its Equity Petal, which was introduced in 2.0 yet was admittedly one of our less developed categories until now. With the integration of JUST™ and a groundbreaking Equity offset framework, the Equity Petal is now as innovative as the rest of the Challenge.
- With 3.0, the LBC continues to raise the bar with materials transparency through the direct connection with Declare™, our materials 'nutrition label,' and an expanded and updated Red List— our first update on the Red List since 2006. Furniture systems are now included in the Materials Petal.
- 3.0 marks the launch of three new Living Future Exchange offset programs—making it easier for project teams to allocate funds to worthwhile causes and see their donations aggregated for greater effect. Our Exchange programs can be found at **www.living-future.org/exchange**
- Further refinements of market-based exceptions throughout the program—including the landmark banning of fluorescent lighting in all but a few applications—the first time a green building program has taken such an important step forward in reducing materials toxicity in the lighting industry.

All in all we, believe 3.0—now part of a larger and more holistic vision of a LIVING FUTURE—is a big step forward. We look forward to learning and getting feedback from all our active project teams and ambassadors worldwide as we immediately begin work on version 3.1.

Imperative omitted from Typology

Solutions beyond project footprint are permissible

The 20 Imperatives of the Living Building Challenge: Follow down the column associated with each Typology to see which Imperatives apply.

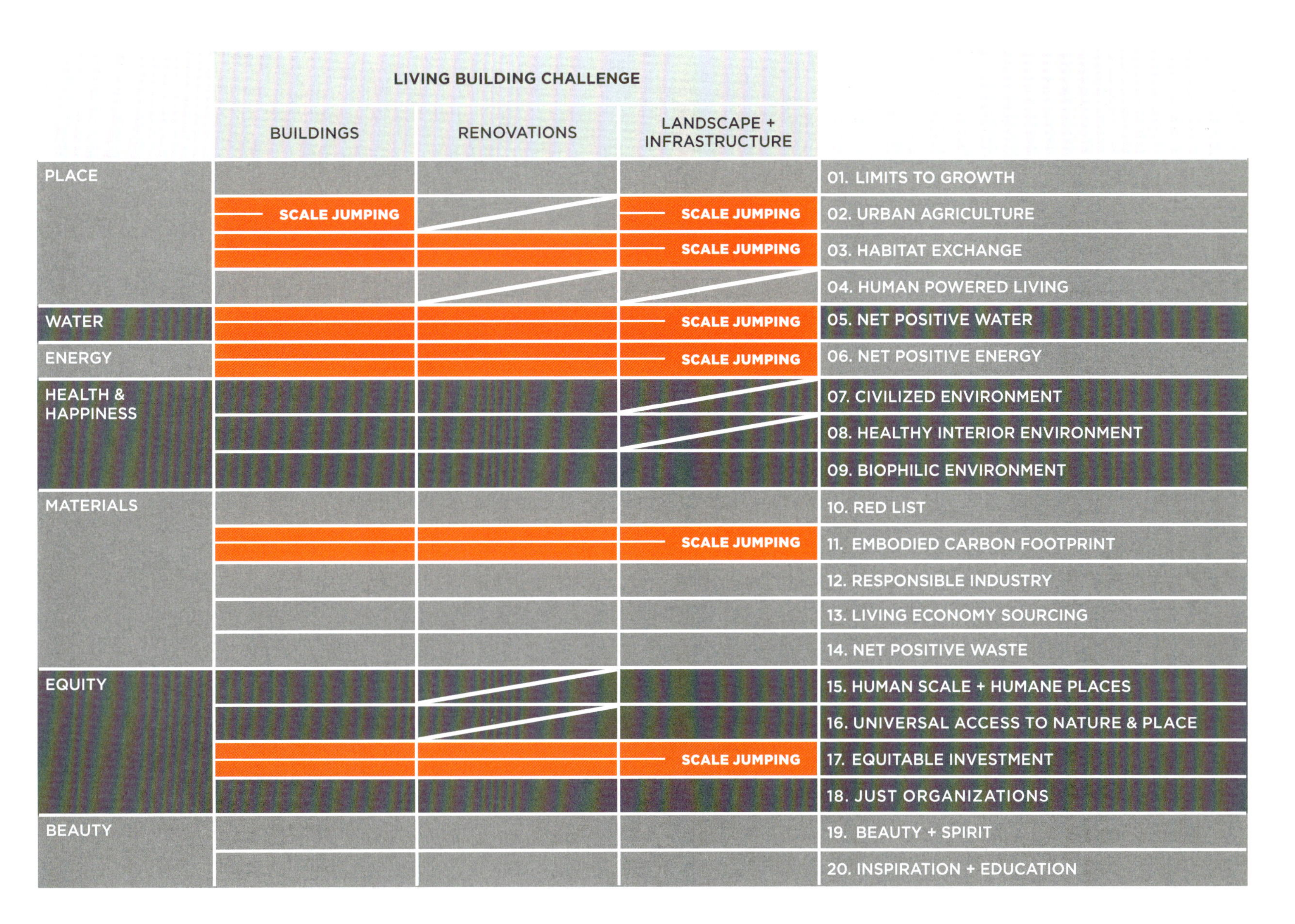

	LIVING BUILDING CHALLENGE			
	BUILDINGS	RENOVATIONS	LANDSCAPE + INFRASTRUCTURE	
PLACE				01. LIMITS TO GROWTH
	SCALE JUMPING		SCALE JUMPING	02. URBAN AGRICULTURE
			SCALE JUMPING	03. HABITAT EXCHANGE
				04. HUMAN POWERED LIVING
WATER			SCALE JUMPING	05. NET POSITIVE WATER
ENERGY			SCALE JUMPING	06. NET POSITIVE ENERGY
HEALTH & HAPPINESS				07. CIVILIZED ENVIRONMENT
				08. HEALTHY INTERIOR ENVIRONMENT
				09. BIOPHILIC ENVIRONMENT
MATERIALS				10. RED LIST
			SCALE JUMPING	11. EMBODIED CARBON FOOTPRINT
				12. RESPONSIBLE INDUSTRY
				13. LIVING ECONOMY SOURCING
				14. NET POSITIVE WASTE
EQUITY				15. HUMAN SCALE + HUMANE PLACES
				16. UNIVERSAL ACCESS TO NATURE & PLACE
			SCALE JUMPING	17. EQUITABLE INVESTMENT
				18. JUST ORGANIZATIONS
BEAUTY				19. BEAUTY + SPIRIT
				20. INSPIRATION + EDUCATION

PLACE

PLACE

RESTORING A HEALTHY INTERRELATIONSHIP WITH NATURE

SCALE JUMPING PERMITTED FOR URBAN AGRICULTURE (IMPERATIVE 02) AND HABITAT EXCHANGE (IMPERATIVE 03)

PETAL INTENT

The intent of the Place Petal is to realign how people understand and relate to the natural environment that sustains us. The human built environment must reconnect with the deep story of place and the unique characteristics found in every community so that story can be honored, protected and enhanced. The Place Petal clearly articulates where it is acceptable for people to build, how to protect and restore a place once it has been developed, and how to encourage the creation of communities that are once again based on the pedestrian rather than the automobile. In turn, these communities need to be supported by a web of local and regional agriculture, since no truly sustainable community can rely on globally sourced food production.

The continued spread of sprawl development and the vastly increasing number of global megapolises threatens the few wild places that remain. The decentralized nature of our communities impedes our capacity to feed ourselves in a sustainable way and also increases transportation impacts and pollution. The overly dense urban centers in turn crowd out healthy natural systems, isolating culture from a sense of place. As prime land for construction diminishes, more development tends to occur in sensitive areas that are easily harmed or destroyed. Invasive species threaten ecosystems, which are already weakened by the constant pressure of existing human developments. Automobiles, often used as single occupancy vehicles, have become integral to our communities when we should depend on "people power" —walking and bicycling—as the primary mode of travel, and supplement it with shared transit.

IDEAL CONDITIONS + CURRENT LIMITATIONS

The Living Building Challenge envisions a moratorium on the seemingly never-ending growth outward, and a focus instead on compact, connected communities with healthy rather than inhumane levels of density—inherently conserving the natural resources that support human health and the farmlands that feed us, while also inviting natural systems back into the daily fabric of our lives. As previously disturbed areas are restored, the trend is reversed and nature's functions are invited back into a healthy interface with the built environment.

Human behavior and attitudes are the most significant barriers to transforming our surroundings. There is a frontier mentality that seems to encourage people to keep pursuing the next open territory and to value the untouched site more than the secondhand site. Humanity is territorial by nature, and we tend to view our impacts through a narrow lens. It is not unusual for us to encourage unhealthy solutions, so long as they are "not in my backyard" and allow us the social stature to "keep up with the Joneses." We must erase the taboo associated with certain forms of transit and abandoned industrial and commercial facilities, and we must once again give our regard to the many others that cohabit the earth with us.

PLACE

LIMITS TO GROWTH

Projects may only be built on greyfields or brownfields: previously developed[6] sites that are not classified as on or adjacent to any of the following sensitive ecological habitats[7]:

- Wetlands: maintain at least 15 meters, and up to 70 meters of separation
- Primary dunes: maintain at least 40 meters of separation
- Old-growth forest: maintain at least 60 meters of separation
- Virgin prairie: maintain at least 30 meters of separation
- Prime farmland
- Within the 100-year flood plain

Project teams must document site conditions prior to the start of work. On-site landscape must be designed so that as it matures and evolves it increasingly emulates the functionality of indigenous ecosystems with regard to density, biodiversity, plant succession, water use, and nutrient needs. It shall also provide wildlife and avian habitat appropriate to the project's transect through the use of native and naturalized plants and topsoil. No petrochemical fertilizers or pesticides can be used for the operation and maintenance of the on-site landscape.

6 Sites that qualify must have been altered from a greenfield prior to December 31, 2007.
7 Refer to the Place Petal Handbook for clarifications and exceptions. There are cases when building on a greenfield or a sensitive ecological habitat is allowed based on project type, Transect or other conditions.

PLACE

URBAN AGRICULTURE

IMPERATIVE

The project must integrate opportunities for agriculture appropriate to its scale and density using the Floor Area Ratio (FAR) as a basis for calculation. The table below outlines the mandatory agricultural requirements for all projects. Single-family homes must also demonstrate the capacity to store at least a two-week supply of food.[8]

PERCENT OF PROJECT AREA FOR FOOD PRODUCTION

Project F.A.R.	Minimum Percent Required
< 0.05	80%
0.05 ≤ 0.09	50%
0.10 ≤ 0.24	35%
0.25 ≤ 0.49	30%
0.5 ≤ 0.74	25%
0.75 ≤ 0.99	20%
1.0 ≤ 1.49	15%
1.5 ≤ 1.99	10%
2.0 ≤ 2.99	5%
> 3.0	1%

8 Refer to the Place Petal Handbook for clarifications such as acceptable urban agriculture practices, area calculation information as well as exceptions by Transect.

PLACE

HABITAT EXCHANGE

For each hectare of development, an equal amount of land away from the project site must be set aside in perpetuity through the Institute's Living Future Habitat Exchange Program[9] or an approved Land Trust organization.[10] The minimum offset amount is 0.4 hectare.

9 ILFI now operates a Habitat Exchange Program in cooperation with conservation organizations. For more information visit **www.living-future.org/exchange.**

10 Refer to the Place Petal Handbook for clarifications such as information about Land Trusts as well as exceptions.

PLACE

HUMAN POWERED LIVING

Each new project should contribute toward the creation of walkable, pedestrian-oriented communities and must not lower the density of the existing site. Teams must evaluate the potential for a project to enhance the ability of a community to support a human powered lifestyle, and provide a mobility plan, which addresses the interior and exterior of the project and demonstrates at a minimum the following:

ALL PROJECTS:

- Secure, weather protected storage for human powered vehicles that provide facilities to encourage biking.[11]
- Consideration and enhancement of pedestrian routes, including weather protection on street frontages.
- Promotion of the use of stairs over elevators through interior layout and quality of stairways.
- Advocacy in the community to facilitate the uptake of human powered transportation.

PROJECTS IN TRANSECTS L4-L6 MUST ALSO PROVIDE:

- A transit subsidy for all occupants of the building (if owner occupied) or a requirement for tenant employers to provide such a subsidy.
- Showers and changing facilities that can be accessed by all occupants of the building.
- At least one electric vehicle charging station.

SINGLE FAMILY HOMES (ALL TRANSECTS):

An assessment of how the residents can reduce their transportation impact through car sharing, use of public transportation, alternative fueled vehicles, or bicycles is required.

11 Bike storage is recommended for 15% of occupants; teams should consider the occupancy type and location of the project.

WATER

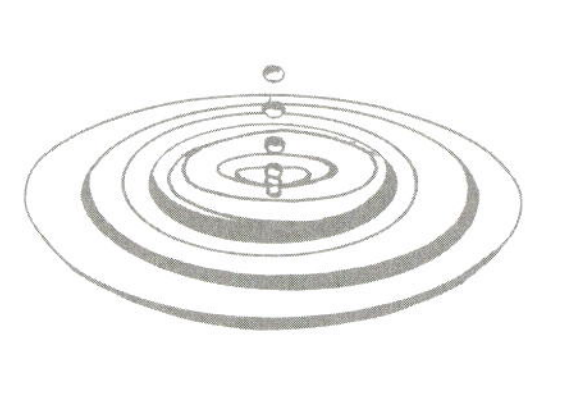

WATER

CREATING DEVELOPMENTS THAT OPERATE WITHIN THE WATER BALANCE OF A GIVEN PLACE AND CLIMATE

PETAL INTENT

The intent of the Water Petal is to realign how people use water and to redefine 'waste' in the built environment, so that water is respected as a precious resource.

Scarcity of potable water is quickly becoming a serious issue as many countries around the world face severe shortages and compromised water quality. Even regions that have avoided the majority of these problems to date due to a historical presence of abundant fresh water are at risk: the impacts of climate change, highly unsustainable water use patterns, and the continued drawdown of major aquifers portend significant problems ahead.

IDEAL CONDITIONS AND CURRENT LIMITATIONS

The Living Building Challenge envisions a future whereby all developments are configured based on the carrying capacity of the site: harvesting sufficient water to meet the needs of a given population while respecting the natural hydrology of the land, the water needs of the ecosystem the site inhabits, and those of its neighbors. Indeed, water can be used and purified and then used again—and the cycle repeats.

Currently, such practices are often illegal due to health, land use and building code regulations (or because of the undemocratic ownership of water rights) that arose precisely because people were not properly safeguarding the quality of their water. Therefore, reaching the ideal for water use means challenging outdated attitudes and technology with decentralized site- or district-level solutions that are appropriately scaled, elegant and efficient.

SCALE JUMPING PERMITTED FOR NET POSITIVE WATER (IMPERATIVE 05)

WATER

NET POSITIVE WATER

Project water use and release must work in harmony with the natural water flows of the site and its surroundings. One hundred percent of the project's water needs must be supplied by captured precipitation or other natural closed loop water systems,[12] and/or by re-cycling used project water, and must be purified as needed without the use of chemicals.

All stormwater and water discharge, including grey and black water, must be treated onsite and managed either through re-use, a closed loop system, or infiltration. Excess stormwater can be released onto adjacent sites under certain conditions.

12 Refer to the Water Petal Handbook for clarifications and exceptions, such as allowances for a municipal potable water use connection if required by local heath regulations.

Omega Institute, Rhinebeck, NY
Living Certification - Living Building Challenge 1.3
Photo: Farshid Assassi / Courtesy: BNIM Architects

ENERGY

Rooftop Solar Array at The Bullitt Center
Seattle, WA
Photo: Benjamin Benschneider

ENERGY

RELYING ONLY ON CURRENT SOLAR INCOME

PETAL INTENT

The intent of the Energy Petal is to signal a new age of design, wherein the built environment relies solely on renewable forms of energy and operates year round in a safe, pollution-free manner. In addition, it aims to prioritize reductions and optimization before technological solutions are applied to eliminate wasteful spending—of energy, resources, and dollars. The majority of energy generated today is from highly polluting and often politically destabilizing sources including coal, gas, oil and nuclear power. Large-scale hydro, while inherently cleaner, results in widespread damage to ecosystems. Burning wood, trash or pellets releases particulates and carbon dioxide (CO_2) into the atmosphere and often strains local supplies of sustainably harvested biomass while robbing the soil of much-needed nutrient recycling. The effects of these energy sources on regional and planetary health are becoming increasingly evident through climate change, the most worrisome major global trend attributed to human activity.

IDEAL CONDITIONS AND CURRENT LIMITATIONS

The Living Building Challenge envisions a safe, reliable and decentralized power grid, powered entirely by renewable energy, supplied to incredibly efficient buildings and infrastructure without the negative externalities associated with combustion or fission.

Although there has been considerable progress made to advance renewable energy technologies, there is still a need for a greater efficiency from these systems and for new, cleaner ways to store the energy they generate. These, together with the current cost of the systems available, are the major limitations to reaching our goals.

SCALE JUMPING PERMITTED FOR NET POSITIVE ENERGY (IMPERATIVE 06)

ENERGY

NET POSITIVE ENERGY

One hundred and five percent of the project's energy needs must be supplied by on-site renewable energy on a net annual basis, without the use of on-site combustion.[13] Projects must provide on-site energy storage for resiliency.[14]

13 Refer to the Energy Petal Handbook for a list of renewable energy systems, clarifications and exceptions.

14 Projects must demonstrate that sufficient back-up battery power be installed for emergency lighting (at least 10 percent of lighting load) and refrigeration use for up to one week for greater resiliency.

Solar array at The Hawaii Preparatory Academy Energy Lab, Kamuela, HI
Living Certification - Living Building Challenge 1.3
Photo: Matthew Millman Photography / Courtesy: Flansburgh Architects

Okanagan College, Kelowna, BC
Courtesy: CEI Architecture

HEALTH & HAPPINESS

HEALTH & HAPPINESS

CREATING ENVIRONMENTS THAT OPTIMIZE PHYSICAL AND PSYCHOLOGICAL HEALTH AND WELL BEING

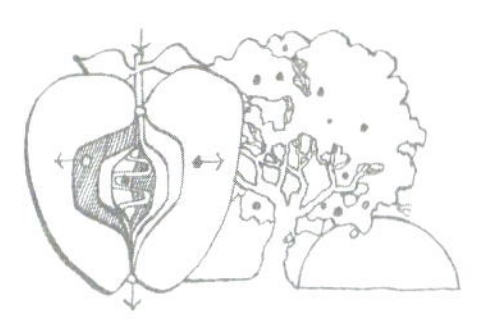

PETAL INTENT

The intent of the Health and Happiness Petal is to focus on the most important environmental conditions that must be present to create robust, healthy spaces, rather than to address all of the potential ways that an interior environment could be compromised.

Many developments provide substandard conditions for health and productivity and human potential is greatly diminished in these places. By focusing attention on the major pathways of health we create environments designed to optimize our well-being.

IDEAL CONDITIONS AND CURRENT LIMITATIONS

The Living Building Challenge envisions a nourishing, highly productive and healthy built environment. However, even best available solutions require acceptance and engagement by the project occupants and project owner. It is difficult to ensure that developments will remain healthy over time, since environmental conditions such as air quality, thermal control, and visual comfort can easily be compromised in numerous ways. It can also be complicated to ensure optimal conditions due to the unpredictable nature of how people operate and maintain their indoor spaces.

HEALTH & HAPPINESS

CIVILIZED ENVIRONMENT

Every regularly occupied space must have operable windows that provide access to fresh air and daylight.[15]

15 Refer to the Health Petal Handbook for clarifications, exceptions and information regarding minimum requirements for windows.

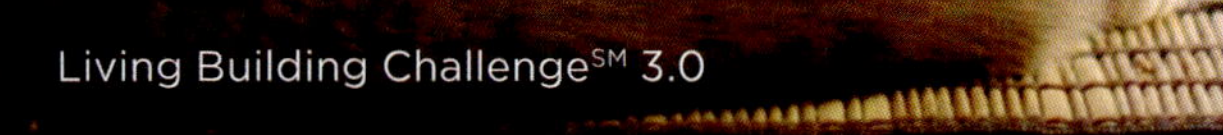

HEALTH & HAPPINESS

HEALTHY INTERIOR ENVIRONMENT

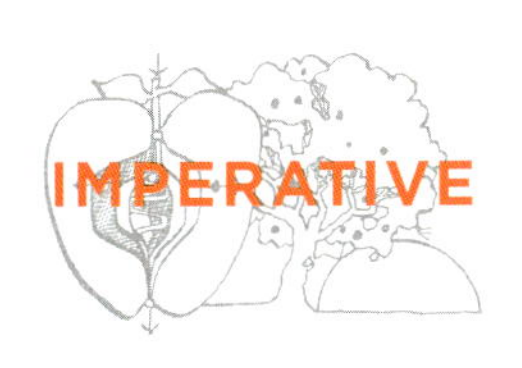

To promote good indoor air quality, a project must create a Healthy Indoor Environment Plan that explains how the project will achieve an exemplary indoor environment and achieve the following:

- Compliance with the current version of ASHRAE 62, or international equivalent
- Smoking must be prohibited within the project boundary
- Results from an Indoor Air Quality test before and nine months after occupancy[16]
- Compliance with the CDPH Standard Method v1.1-2010 (or international equivalent) for all interior building products that have the potential to emit Volatile Organic Compounds[17]
- Dedicated exhaust systems for kitchens, bathrooms, and janitorial areas
- An entry approach that addresses particulate reduction tracked in through shoes[18]
- An outline of a cleaning protocol that uses cleaning products that comply with the EPA Design for Environment label (or international equivalent[19])

16 Testing protocols must be consistent with the United States Environmental Protection Agency Compendium of Methods for the Determination, or International equivalent. Refer to the Health Petal Handbook for the required Air Quality Conditions.
17 California Department of Public Health. Products not regulated by CDHP do not need to comply.
18 Refer to the Health Petal Handbook for the specifics of approved entry strategies.
19 www.epa.gov/dfe

NRDC Midwest Office, Chicago, IL
Petal Certification
Courtesy: Studio Gang Architects

HEALTH & HAPPINESS

BIOPHILIC ENVIRONMENT

IMPERATIVE 09

The project must be designed to include elements that nurture the innate human/ nature connection. Each project team must engage in a minimum of one all-day exploration of the biophilic design potential for the project. The exploration must result in a biophilic framework and plan for the project that outlines the following:[20]

- How the project will be transformed by deliberately incorporating nature through Environmental Features, Light and Space, and Natural Shapes and Forms
- How the project will be transformed by deliberately incorporating nature's patterns through Natural Patterns and Processes and Evolved Human-Nature Relationships
- How the project will be uniquely connected to the place, climate and culture through Place-based Relationships
- The provision of sufficient and frequent human-nature interactions in both the interior and exterior of the project to connect the majority of occupants with nature directly

The plan must contain methods for tracking biophilia at each design phase. The plan should include historical, cultural, ecological, and climatic studies that thoroughly examine the site and context for the project.

20 Each of the Biophilic Design Elements outlined on Table 1-1, Page 15 of Biophilic Design: The Theory, Science, and Practice of Bringing Buildings to Life by Stephen R. Kellert, Judith H. Heerwagen, and Martin L. Mador should be used as a reference.

Omega Institute, Rhinebeck, NY
Living Certification - Living Building Challenge 1.3
Photo: Farshid Assassi / Courtesy: BNIM Architects

Phipps Conservatory and Botanical Gardens, Pittsburgh, PA
Net Zero Energy Building Certification
Photo: Denmarsh Photography, Inc.

MATERIALS

UniverCity Childcare Centre
Burnaby, BC
Courtesy: space2place

MATERIALS

ENDORSING PRODUCTS THAT ARE SAFE FOR ALL SPECIES THROUGH TIME

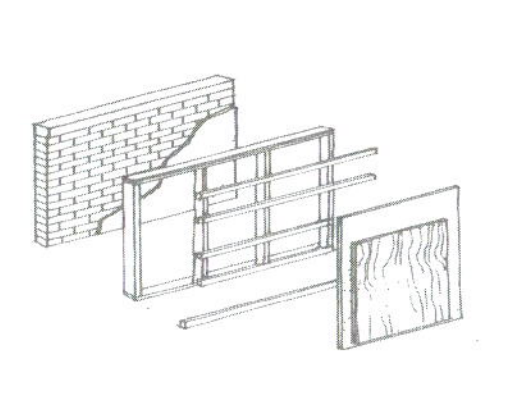

PETAL INTENT

The intent of the Materials Petal is to help create a materials economy that is non-toxic, ecologically regenerative, transparent and socially equitable. Throughout their life cycle, building materials are responsible for many adverse environmental issues, including personal illness, habitat and species loss, pollution, and resource depletion. The Imperatives in this section aim to remove the worst known offending materials and practices and drive business towards a truly responsible materials economy. When impacts can be reduced but not eliminated, there is an obligation not only to offset the damaging consequences associated with the construction process, but also to strive for corrections in the industry itself. At the present time it is impossible to gauge the true environmental impact and toxicity of the built environment due to a lack of product-level information, although the Living Building Challenge continues to shine a light on the need for transformative industrial practices.

IDEAL CONDITIONS + CURRENT LIMITATIONS

The Living Building Challenge envisions a future where all materials in the built environment are regenerative and have no negative impact on human and ecosystem health. The precautionary principle guides all materials decisions when impacts are unclear.

There are significant limitations to achieving the ideal for the materials realm. Product specification and purchase has far-reaching impacts, and although consumers are starting to weigh these in parallel with other more conventional attributes, such as aesthetics, function and cost, the biggest shortcoming is due to the market itself. While there are a huge number of "green" products for sale, there is also a shortage of good, publicly available data that backs up manufacturer claims and provides consumers with the ability to make conscious, informed choices. Transparency is vital; as a global community, the only way we can transform into a truly sustainable society is through open communication and honest information sharing, yet many manufacturers are wary of sharing trade secrets that afford them a competitive advantage, and make proprietary claims about specific product contents.

Declare, the Institute's ingredients label for building products, is a publicly accessible label and online database with an official connection to the Materials Petal. Not only does Declare contribute to the overt methodology for removing a temporary exception, it also provides a forum for sharing the information compiled by a project team as part of their documentation requirements for certification.

declareproducts.com

SCALE JUMPING PERMITTED FOR EMBODIED CARBON FOOTPRINT (IMPERATIVE 11)

The Hawaii Preparatory Academy Energy Lab, Kamuela, HI
Living Certification - Living Building Challenge 1.3
Photo: Matthew Millman Photography / Courtesy: Flansburgh Architects

MATERIALS

RED LIST

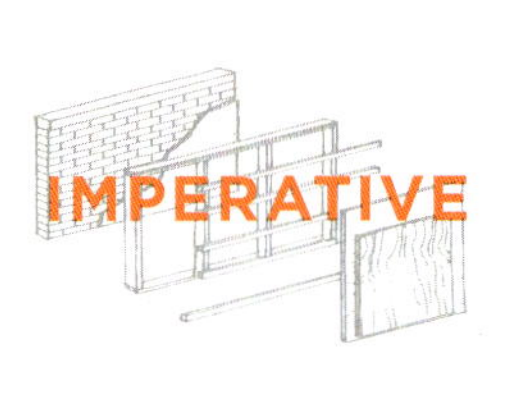

There are temporary exceptions for numerous Red List items due to current limitations in the materials economy. Refer to the Materials Petal Handbook for complete and up-to-date listings.

The project cannot contain any of the following Red List materials or chemicals:[21]

- Alkylphenols
- Asbestos
- Bisphenol A (BPA)
- Cadmium
- Chlorinated Polyethylene and Chlorosulfonated Polyethlene
- Chlorobenzenes
- Chlorofluorocarbons (CFCs) and Hydrochlorofluorocarbons (HCFCs)
- Chloroprene (Neoprene)
- Chromium VI
- Chlorinated Polyvinyl Chloride (CPVC)
- Formaldehyde (added)
- Halogenated Flame Retardants (HFRs)
- Lead (added)
- Mercury
- Polychlorinated Biphenyls (PCBs)
- Perfluorinated Compounds (PFCs)
- Phthalates
- Polyvinyl Chloride (PVC)
- Polyvinylidene Chloride (PVDC)
- Short Chain Chlorinated Paraffins
- Wood treatments containing Creosote, Arsenic or Pentachlorophenol
- Volatile Organic Compounds (VOCs) in wet applied products[23]

21 A link to the list of CAS Registry Numbers that correspond with each Red List item is available in the Materials Petal Handbook.

22 Wet applied products (coatings, adhesives and sealants) must have VOC levels below the South Coast Air Quality Management District (SCAQMD) Rule 1168 for Adhesives and Sealants or the CARB 2007 Suggested Control Measure (SCM) for Architectural Coatings as applicable. Containers of sealants and adhesives with capacity of 16 ounces or less must comply with applicable category limits in the California Air Resources Board (CARB) Regulation for Reducing Emissions from Consumer Products.

MATERIALS

EMBODIED CARBON FOOTPRINT

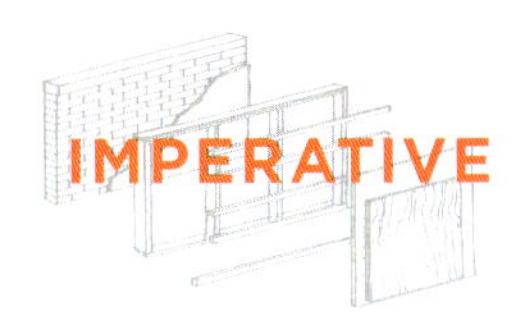

IMPERATIVE 11

The project must account for the total embodied carbon (tCO_2e) impact from its construction through a one-time carbon offset in the Institute's new Living Future Carbon Exchange or an approved carbon offset provider.[23]

23 Refer to the Materials Petal Handbook for approved carbon offset programs, clarifications and exceptions.

Omega Institute, Rhinebeck, NY
Living Certification - Living Building Challenge 1.3
Photo: Farshid Assassi / Courtesy: BNIM Architects

MATERIALS

RESPONSIBLE INDUSTRY

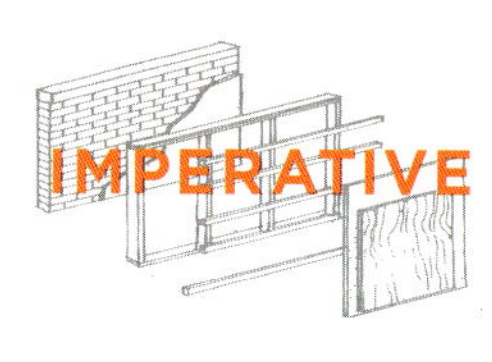

The project must advocate for the creation and adoption of third-party certified standards for sustainable resource extraction and fair labor practices. Applicable raw materials include stone and rock, metal, minerals, and timber.

For timber, all wood must be certified to Forest Stewardship Council (FSC)[24] 100% labeling standards, from salvaged sources, or from the intentional harvest of timber onsite for the purpose of clearing the area for construction or restoring/maintaining the continued ecological function of the onsite bionetwork.

All projects must use—at a minimum—1 Declare product for every 500 square meters of project area and must send Declare program information[25] to at least 10 manufacturers not currently using Declare.

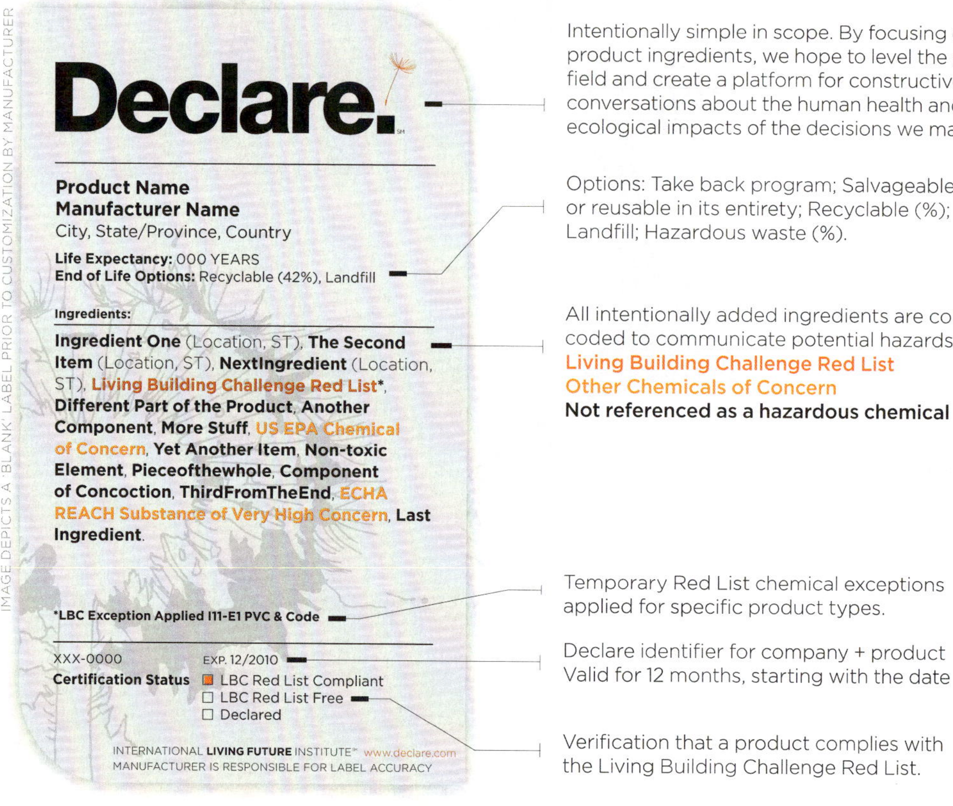

IMAGE DEPICTS A 'BLANK' LABEL PRIOR TO CUSTOMIZATION BY MANUFACTURER

24 Refer to the Materials Petal Handbook for a full list of exceptions such as an exception for wood in existing buildings undergoing renovation.

25 www.declareproducts.com

MATERIALS

LIVING ECONOMY SOURCING

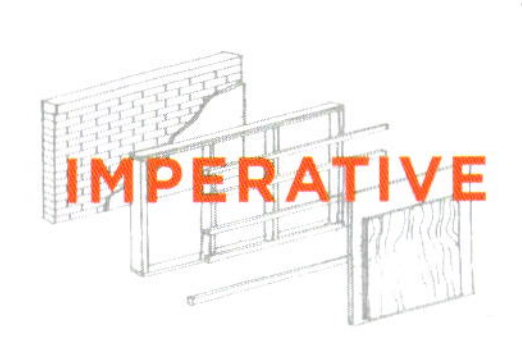

The project must incorporate place-based solutions and contribute to the expansion of a regional economy rooted in sustainable practices, products and services.

Manufacturer location for materials and services must adhere to the following restrictions:

- 20% or more of materials construction budget[26] must come from within 500 km of construction site.
- An additional 30% of materials construction budget must come from within 1000 km of the construction site or closer.
- An additional 25% of materials construction budget must come from within 5000 km of the construction site.
- 25% of materials may be sourced from any location.
- Consultants must come from within 2500 km of the project location.[27]

26 Materials construction budget is defined as all material costs and excludes labor, soft costs and land. Declare products and salvaged materials may be counted at twice their value. Certain natural building materials may include labor cost in their calculation. Refer to the Materials Petal Handbook for more information.

27 There is a temporary exception for specialty consultants and subcontractors, who may travel up to 5,000 km. Refer to the Materials Petal Handbook for additional exceptions.

Painters Hall
Courtesy: Pringle Creek Community

MATERIALS

NET POSITIVE WASTE

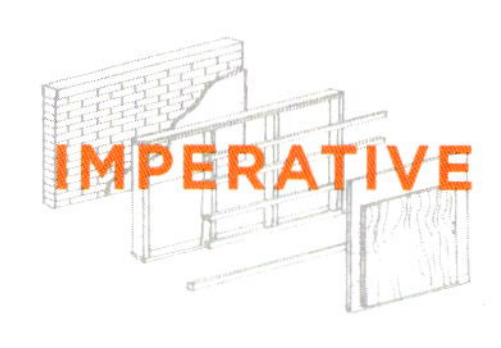

The project team must strive to reduce or eliminate the production of waste during design, construction, operation, and end of life in order to conserve natural resources and to find ways to integrate waste back into either an industrial loop or natural nutrient loop.[28]

All Projects must feature at least one salvaged material per 500 square meters or be an adaptive reuse of an existing structure.

The project team must create a Material Conservation Management Plan that explains how the project optimizes materials in each of the following phases:

- Design Phase, including the consideration of appropriate durability in product specification
- Construction Phase, including product optimization and collection of wasted materials
- Operation Phase, including a collection plan for consumables and durables
- End of Life Phase, including a plan for adaptable reuse and deconstruction

During construction, the project team must divert wasted material to the following levels:

MATERIAL	MINIMUM DIVERTED/WEIGHT
Metal	99%
Paper & Cardboard	99%
Soil & Biomass	100%
Rigid foam, Carpet & Insulation	95%
All others - combined weighted average[29]	90%

For all project types, there must be dedicated infrastructure for the collection of recyclables and compostable food scraps.

A project that is located on a site with existing infrastructure must complete a pre-building audit that inventories available materials and assemblies for reuse or donation.

28 Refer to the Materials Petal Handbook for calculation details, clarifications and exceptions.
29 Hazardous materials in demolition waste, such as lead-based paint, asbestos, and polychlorinated biphenyls (PCBs), are exempt from percentage calculations.

VanDusen Botanical Garden
Visitor Center, Vancouver, BC
Photo: Nic Lehoux / Courtesy: Perkins+Will

EQUITY

The Bullitt Center, Seattle, WA
Photo: Benjamin Benschneider

EQUITY

SUPPORTING A JUST, EQUITABLE WORLD

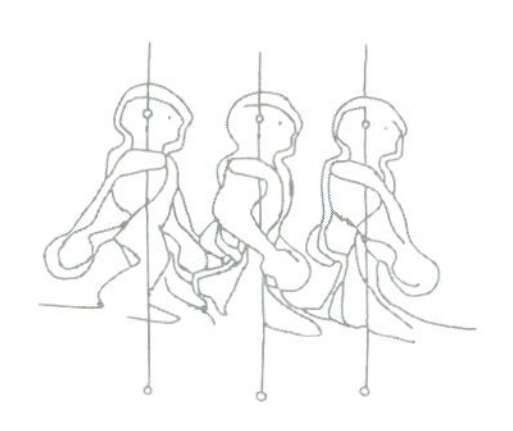

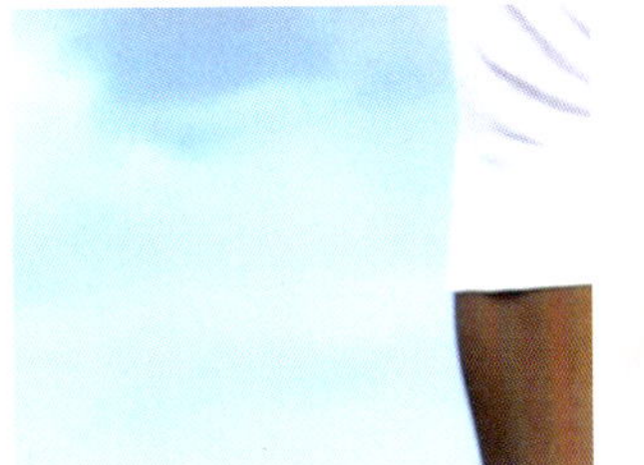

SCALE JUMPING PERMITTED

PETAL INTENT

The intent of the Equity Petal is to transform developments to foster a true, inclusive sense of community that is just and equitable regardless of an individual's background, age, class, race, gender or sexual orientation. A society that embraces all sectors of humanity and allows the dignity of equal access and fair treatment is a civilization in the best position to make decisions that protect and restore the natural environment that sustains all of us.

There is a disturbing trend toward privatizing infrastructure and creating polarized attitudes of 'us' vs. 'them'—allowing only those of a certain economic or cultural background to participate fully in community life. Although opposite on the spectrum, enclaves for the wealthy are only one step removed from the racial and ethnic ghettos that continue to plague our neighborhoods. A subset of this trend is the notion that individuals can own access to nature itself, by privatizing admittance to waterways, beaches and other wilderness areas, cutting off most people from the few pristine environmental places that remain. Only by realizing that we are indeed all in this together can the greatest environmental and social problems be addressed.

We need to aggressively challenge the notion that property ownership somehow implies that we can do whatever we like, even externalize the negative environmental impacts of our actions onto others.

For example, consider these situations: when a polluting factory is placed next to a residential community, the environmental burdens of its operation are placed on the individuals who live in those houses. The factory is diminishing its neighbors' rights to clean air, water and soil. When a building towers over another structure, its shadow diminishes that structure's ability to generate clean and renewable energy, thereby impeding the rights to energy independence. We all deserve access to sunlight and clean air, water and soil.

We need to prioritize the concept of "citizen" above that of "consumer." Equity implies the creation of communities that provide universal access to people with disabilities, and allow people who can't afford expensive forms of transportation to fully participate in the major elements of society. Indeed, most projects in the built environment greatly outlive the original owner or developer—society inherits

continued >>

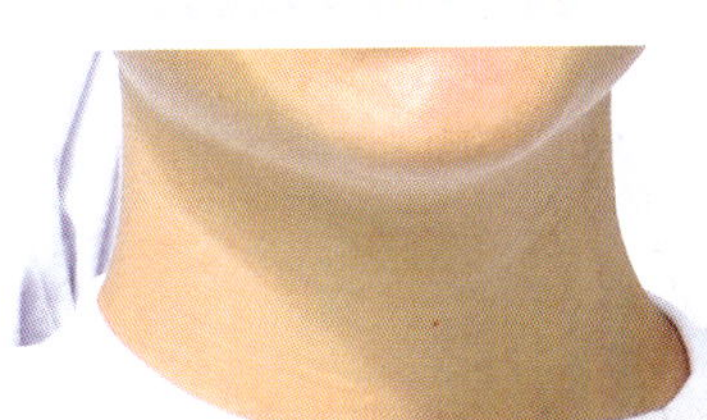

EQUITY

SUPPORTING A JUST, EQUITABLE WORLD

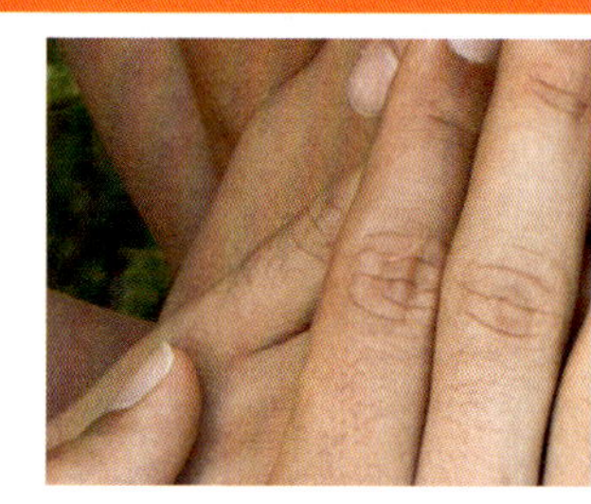

the legacies of bad decisions and good decisions alike. Since the act of building is a considerable environmental impact shared by all, there is an inherent responsibility to ensure that any project provides some public good and does not degrade quality of life. Finally, it is essential that we recognize the business practices and welfare of the people that we support as we design and build our developments.

JUST, the Institute's ingredients label for social justice, is a publicly accessible label and online database with an official connection to the Equity Petal. JUST provides a powerful forum for helping project teams support organizations that share the values of a responsible equitable living future.

justorganizations.org

IDEAL CONDITIONS + CURRENT LIMITATIONS

The Living Building Challenge envisions communities that allow equitable access and treatment to all people regardless of physical abilities, age, or socioeconomic status.

Current limitations to reaching this ideal stem from ingrained cultural attitudes about the rights associated with private ownership and the varying rights of people.

It is necessary to change zoning standards in order to protect the rights of individuals who are 'downstream' of water, air and noise pollution, and who are adversely impacted due to lack of sunlight or exposure to toxins. Past attempts by zoning standards to protect people from particularly egregious pollutants resulted in sterile, single-use areas. A healthy, diverse community is one that encourages multiple functions, and is organized in a way that protects the health of people and the environment.

SCALE JUMPING PERMITTED

EQUITY

HUMAN SCALE AND HUMANE PLACES

The project must be designed to create human-scaled rather than automobile-scaled places so that the experience brings out the best in humanity and promotes culture and interaction.
In context of the character of each Transect, there are specific maximum (and sometimes minimum) requirements for paved areas, street and block design, building scale and signage that contribute to livable places.

The project must follow the following design guidelines:

<table>
<tr><th colspan="2">TRANSECT</th><th>L1</th><th>L2</th><th>L3</th><th>L4</th><th>L5</th><th>L6</th></tr>
<tr><td rowspan="2">Surface Cover</td><td>Maximum dimension of surface parking lot before a separation is required on all four sides
e.g., building, wall, or 3 m wide (minimum) planted median or bioswale</td><td colspan="6">20 m x 30 m</td></tr>
<tr><td>Total area of surface parking lot allowed. All other parking requirements must be handled in structured or underground parking.</td><td>20%</td><td>20%</td><td>20%</td><td>15%</td><td>5%</td><td>0%</td></tr>
<tr><th colspan="2">TRANSECT</th><th>L1</th><th>L2</th><th>L3</th><th>L4</th><th>L5</th><th>L6</th></tr>
<tr><td rowspan="7">Streets + Intersections
Only applicable if adding new streets</td><td>Maximum street width, measured either shoulder-to-shoulder or curb-to-curb</td><td colspan="2">5 m</td><td>7.5 m</td><td>10 m</td><td>15 m</td><td>22.5 m</td></tr>
<tr><td>Maximum street width before driving lanes must be separated by a pedestrian strip and planting median. Additional lanes may be included on the other side of median to a maximum of 22.5 m total width of driving area</td><td colspan="2">Not applicable</td><td colspan="4">15 m</td></tr>
<tr><td>Maximum street width before tree plantings and sidewalks are required on both sides</td><td colspan="2" rowspan="5">Development of this kind is not permitted in a Natural Habitat Preserve or Rural Agricultural Zone</td><td colspan="4">7.5 m</td></tr>
<tr><td>Minimum overall width of sidewalks and planted median</td><td colspan="4">1/3 street width</td></tr>
<tr><td>Maximum distance between trees in furnishing zone and planted median</td><td colspan="4">9 m</td></tr>
<tr><td>Maximum distance between circulation routes
Access way must be 3 m wide minimum to qualify</td><td colspan="2">45 m</td><td colspan="2">60 m</td></tr>
<tr><td>Maximum street block size</td><td colspan="2">60 m x 120 m</td><td colspan="2">120 m x 120 m</td></tr>
<tr><th colspan="2">TRANSECT</th><th>L1</th><th>L2</th><th>L3</th><th>L4</th><th>L5</th><th>L6</th></tr>
<tr><td rowspan="3">Signage</td><td>Number of free-standing signs per development</td><td colspan="6">1</td></tr>
<tr><td>Maximum dimensions of free-standing sign(s)</td><td colspan="2">2 m x 2.5 m</td><td colspan="2">2.5 m x 3 m</td><td colspan="2">3.5 m x 6 m</td></tr>
<tr><td>Maximum elevation of sign's bottom edge above ground</td><td>2 m</td><td>3 m</td><td>6 m</td><td>9 m</td><td>12 m</td><td>12 m or roof-mounted</td></tr>
<tr><th colspan="2">TRANSECT</th><th>L1</th><th>L2</th><th>L3</th><th>L4</th><th>L5</th><th>L6</th></tr>
<tr><td rowspan="3">Proportion</td><td>Maximum Single Family Residence Size</td><td>N/A</td><td colspan="5">425 m²</td></tr>
<tr><td>Maximum distance between façade openings</td><td>N/A</td><td colspan="5">30 m</td></tr>
<tr><td>Maximum footprint for any building with a single use, single owner or single tenant.
Acceptable to provide additional floor area for tenant on upper/lower floor(s)</td><td colspan="6">3750 m²
excludes floor area of atriums, courtyards and daylight shafts</td></tr>
<tr><td rowspan="2">Human Scale</td><td>Provision of places for people to gather and connect internally and/or with the neighborhood.</td><td>1</td><td>1</td><td colspan="4">One every 1000 m² (10,760sf)</td></tr>
<tr><td>Provision of elements along the project edge which support the human scale of the larger neighborhood, such as seat walls, art, displays, or pocket parks. Single Family residences are excluded</td><td>1</td><td>1</td><td colspan="4">One every 4000 m² (43,000sf)</td></tr>
</table>

UniverCity Childcare Centre, Burnaby, BC
Photo: Martin Tessler

EQUITY

UNIVERSAL ACCESS TO NATURE & PLACE

All primary transportation, roads and non-building infrastructure that are considered externally focused must be equally accessible[30] to all members of the public regardless of background, age and socioeconomic class—including the homeless—with reasonable steps taken to ensure that all people can benefit from the project's creation.

For any project (except single family residential) located in Transect L3-L6, the public realm must be provided for and enhanced through design measures and features such as street furniture, public art, gardens and benches that are accessible to all members of society.

Access for those with physical disabilities must be safeguarded through designs meeting the Americans with Disabilities Act (ADA) and Architectural Barriers Act (ABA) Accessibility Guidelines.[31]

continued >>

30 Refer to the Equity Petal Handbook for a complete list of applicable infrastructure and exceptions that address issues of safety.
31 Refer to the Equity Petal Handbook for specific exceptions, such as those for private residences and historic structures. Complete ADA and ABA Accessibility Guidelines are available online: www.access-board.gov/adaag/about

Science Education at Phipps Conservatory and Botanical Gardens, Pittsburgh, PA
Net Zero Energy Building Certification
Photo: Cory Doman

EQUITY

UNIVERSAL ACCESS TO NATURE & PLACE

IMPERATIVE 16

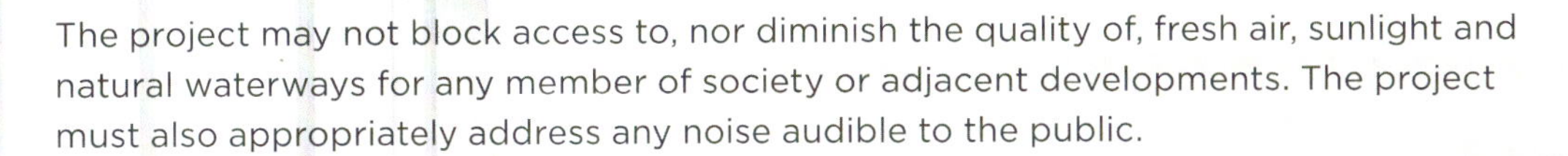

The project may not block access to, nor diminish the quality of, fresh air, sunlight and natural waterways for any member of society or adjacent developments. The project must also appropriately address any noise audible to the public.

- **Fresh Air:** The project must protect adjacent property from any noxious emissions that would compromise its ability to use natural ventilation. All operational emissions must be free of Red List items, persistent bioaccumulative toxicants, and known or suspect carcinogenic, mutagenic and reprotoxic chemicals.

- **Sunlight:** The project may not block sunlight to adjacent building façades and rooftops above a maximum height allotted for the Transect.[32]

  The project may not shade the roof of a development with which it shares a party wall, unless the adjoining development was built to a lesser density than acceptable for the Transect.[33]

- **Natural Waterways:** The project may not restrict access[34] to the edge of any natural waterway, except where such access can be proven to be a hazard to public safety or would severely compromise the function of the development.[35] No project may assume ownership of water contained in these bodies or compromise the quality or quantity that flows downstream. If the project's boundary is more than sixty meters long parallel to the edge of the waterway, it must incorporate and maintain an access path to the waterway from the most convenient public right-of-way.[36]

32 Detailed exceptions relating to transects are in the Equity Petal Handbook
33 This corresponds to a neighboring building that is at least two stories in L2-L3; four stories in L4; eight stories in L5; and sixteen stories in L6.
34 Public access throughway must allow approach to waterway from land for pedestrians and bicyclists, and from the water via boat. No infrastructure to support any water-based transport is required.
35 For example, a working dock or marina might need to restrict shoreline access for safety reasons. A private residence may not.
36 The easement containing the pathway must be at least three meters wide and allow entry to both pedestrians and bicyclists.

EQUITY

EQUITABLE INVESTMENT

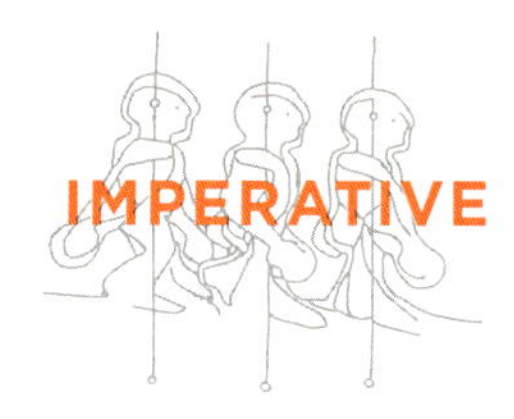

IMPERATIVE 17

For every dollar of total project cost,[37] the development must set aside and donate half a cent or more to a charity[38] of its choosing or contribute to ILFI's Equitable Offset Program, which directly funds renewable infrastructure for charitable enterprises.[39/40]

37 Project cost includes land, soft costs, hard costs and FFE.
38 The Charity must be located in the country of the project and be a registered charity or 501 c 3.
39 Projects may choose to split the offset as desired between multiple charities or ILFI's offset program.
40 Public agencies and charitable organizations are exempt from this requirement.

Global Change Institute at University of Queensland, Brisbane, Australia
Courtesy: HASSELL

EQUITY

JUST ORGANIZATIONS

The project must help create a more JUST, equitable society through the transparent disclosure of the business practices of the major organizations involved. At least one of the following project team members must have a JUST Label for their organization:

- Architect of Record
- MEP Engineer of Record
- Structural Engineer of Record
- Landscape or Interior Architect of Record
- Owner/Developer

Project teams are also required to send JUST program information[41] to at least ten project consultants, sub-consultants or product suppliers as part of ongoing advocacy.

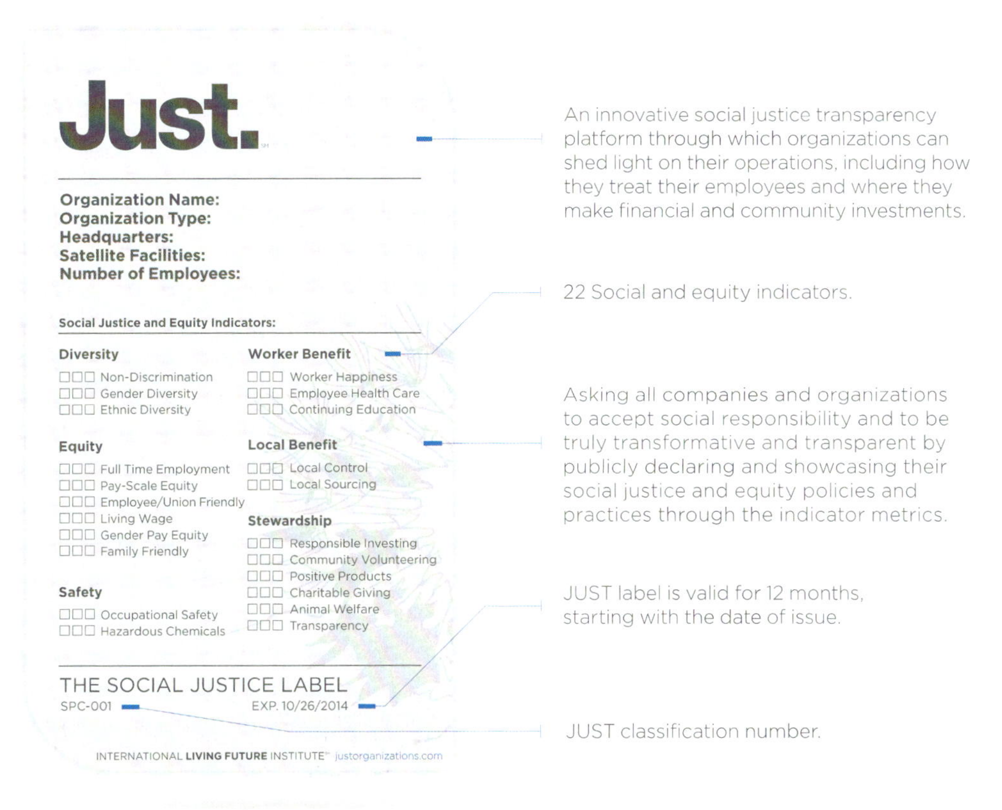

Just.

Organization Name:
Organization Type:
Headquarters:
Satellite Facilities:
Number of Employees:

Social Justice and Equity Indicators:

Diversity
☐☐☐ Non-Discrimination
☐☐☐ Gender Diversity
☐☐☐ Ethnic Diversity

Equity
☐☐☐ Full Time Employment
☐☐☐ Pay-Scale Equity
☐☐☐ Employee/Union Friendly
☐☐☐ Living Wage
☐☐☐ Gender Pay Equity
☐☐☐ Family Friendly

Safety
☐☐☐ Occupational Safety
☐☐☐ Hazardous Chemicals

Worker Benefit
☐☐☐ Worker Happiness
☐☐☐ Employee Health Care
☐☐☐ Continuing Education

Local Benefit
☐☐☐ Local Control
☐☐☐ Local Sourcing

Stewardship
☐☐☐ Responsible Investing
☐☐☐ Community Volunteering
☐☐☐ Positive Products
☐☐☐ Charitable Giving
☐☐☐ Animal Welfare
☐☐☐ Transparency

THE SOCIAL JUSTICE LABEL
SPC-001 EXP. 10/26/2014

INTERNATIONAL **LIVING FUTURE** INSTITUTE℠ justorganizations.com

41 www.justorganizations.com

The Bullitt Center, Seattle, WA
Photo: Benjamin Benschneider

BEAUTY

VanDusen Botanical Garden
Visitor Center, Vancouver, BC
Photo: Nic Lehoux / Courtesy: Perkins+Will

BEAUTY

CELEBRATING DESIGN THAT UPLIFTS THE HUMAN SPIRIT

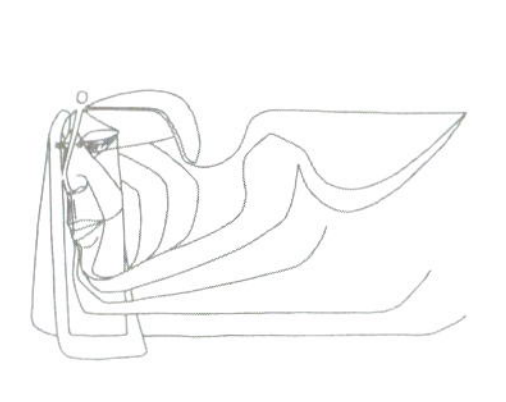

PETAL INTENT

The intent of the Beauty Petal is to recognize the need for beauty as a precursor to caring enough to preserve, conserve and serve the greater good. As a society, we are often surrounded by ugly and inhumane physical environments. If we do not care for our homes, streets, offices and neighborhoods, then why should we extend care outward to our farms, forests and fields? When we accept billboards, parking lots, freeways and strip malls as being aesthetically acceptable, in the same breath we accept clear-cuts, factory farms and strip mines.

IDEAL CONDITIONS AND CURRENT LIMITATIONS

The Living Building Challenge envisions designs that elevate our spirits and inspire us to be better than we currently are. Mandating beauty is, by definition, an impossible task. And yet, the level of discussion and, ultimately, the results are elevated through attempting difficult but critical tasks. In this Petal, the Imperatives are based on genuine efforts, thoughtfully applied. We do not begin to assume we can judge beauty and project our own aesthetic values on others. But we do want to understand people's objectives and know that an effort was made to enrich people's lives with each square meter of construction, on each project. This intentionality of good design and graceful execution must carry forth into a program for educating the public about the environmental qualities of their Living Building Challenge project.

There are no current limitations to this Petal other than our imaginations and what we as a society choose to value.

Green Roof at Phipps Conservatory and Botanical Gardens, Pittsburgh, PA
Net Zero Energy Building Certification
Photo: Paul G. Wiegman

BEAUTY

BEAUTY & SPIRIT

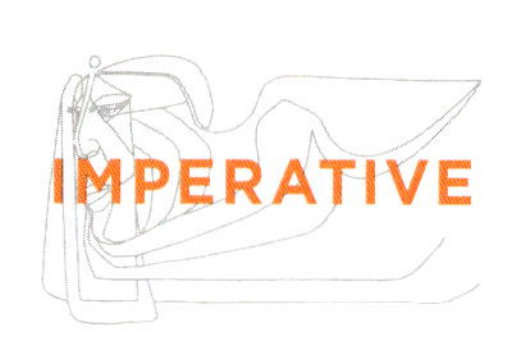

IMPERATIVE 19

The project must contain design features intended solely for human delight and the celebration of culture, spirit and place appropriate to its function and meaningfully integrate public art.

Omega Institute, Rhinebeck, NY
Living Certification - Living Building Challenge 1.3
Photo: Farshid Assassi / Courtesy: BNIM Architects

BEAUTY

INSPIRATION & EDUCATION

Educational materials about the operation and performance of the project must be provided to the public to share successful solutions and to motivate others to make change.

Projects must provide:[42]

- An annual open day for the public.
- An educational web site that shares information about the design, construction, and operation of the project.
- A simple brochure describing the design and environmental features of the project, as well as ways for occupants to optimize project function.
- A copy of the Operations and Maintenance Manual.
- Interpretive signage that teaches visitors and occupants about the project.
- A Living Building Case Study to be posted on the Institute website.

42 Refer to the Beauty and Inspiration Petal Handbook for additional information.

Living wall at Bertschi School, Seattle, WA
Living Certification - Living Building Challenge 2.0
Photo: Benjamin Benschneider

A BRIEF HISTORY OF THE LIVING BUILDING CHALLENGE

The idea for Living Building Challenge emerged in the mid-1990s, during an effort to produce the most advanced sustainable design project in the world: The EpiCenter in Bozeman, Montana. This project was led by Bob Berkebile and Kath Williams and was funded by the National Institute of Standards and Technology. Working with Berkebile at BNIM, Jason F. McLennan guided the research and technology solutions for the EpiCenter—in the process, he also began to conceptualize the requirements for what is now known as a Living Building℠. Following the EpiCenter, Berkebile and McLennan continued to develop the idea and published several related articles.[43]

In 2000, BNIM was hired by the David and Lucile Packard Foundation to examine the economic and environmental implications of a Living Building alongside the various levels of LEED® certification. The findings were presented in a document called the Packard Matrix,[44] which demonstrated that a Living Building was the smartest long-term choice economically, although it carried a hefty first-cost premium. (In 2009, the Institute's Living Building Financial Study proved that first-cost premiums have diminished, and certain building types make immediate financial sense.) More recently, real cost data from completed projects have rounded out the picture, proving that the economic argument for Living Buildings is quite compelling and first-cost premiums modest and diminishing.

In 2005, McLennan began to turn the theoretical idea into a codified standard. He gifted the Living Building Challenge version 1.0 to the Cascadia Green Building Council in August 2006, and three months later the Challenge was formally launched to the public. In 2007, McLennan hired Eden Brukman to direct the ongoing development and international deployment of the Living Building Challenge.

43 Refer to the In The News section of the Institute website to download early publications.
44 www.bnim.com/work/david-and-lucile-packard-foundation-sustainability-report-and-matrix

Together, they authored Living Building Challenge 2.0, rounding out the requirements of the program and demonstrating how to apply the Imperatives to various scales of development and settings.

In response to an increase in global attention and interest, Cascadia founded the International Living Building Institute in 2009 as an umbrella organization for the Living Building Challenge and its auxiliary programs. The Institute certified the first projects in 2010, which changed the green building movement on a fundamental level. Groups around the world reached out to learn more about the Living Building Challenge and to forge formal ties with the Institute, underscoring the truth that people from all parts of the world are looking for hopeful, practical responses to environmental, social and economic difficulties.

At the beginning of 2011, the Institute was renamed the International Living Future Institute, with a mission to lead the transformation to a world that is socially just, culturally rich and ecologically restorative. In 2012, Amanda Sturgeon took over as Director of the Challenge and has led the process to strengthen tools and ease implementation for projects with great success.

As of 2014, over 5 million square feet of LBC projects are underway, representing over a dozen building types in nearly every climate zone on the planet. The ILFI itself moved into a building pursuing Living Certification—The Bullitt Center in Seattle, Washington—in 2013. We are now proud to launch the larger framework of the Living Future Challenge, authored by McLennan and team—of which the Living Building Challenge 3.0 is the Institute's flagship program for deep systemic change. The Institute offers global solutions for lasting sustainability, partners with local communities to create grounded and relevant solutions, and reaches out to individuals to unleash their imagination and innovation.

Phipps Conservatory and Botanical Gardens, Pittsburgh, PA
Net Zero Energy Building Certification
Photo: Denmarsh Photography, Inc.

THE INSTITUTE CONTINUALLY WORKS TO CREATE RESOURCES THAT ADVANCE THE UNDERSTANDING AND IMPLEMENTATION OF THE PRINCIPLES OF THE LIVING BUILDING CHALLENGE, AND WE WANT TO ENSURE THAT ALL ENTHUSIASTS ARE AWARE OF THE VARIOUS WAYS TO LEARN MORE ABOUT AND PARTICIPATE IN THE EVOLUTION OF THE PROGRAM. THIS SECTION LISTS SEVERAL OFFERINGS CREATED BY THE INSTITUTE THAT EXPAND THE ROLE OF THE LIVING BUILDING CHALLENGE BEYOND A FRAMEWORK FOR DEVELOPMENT, TO AN OVERLAY FOR EDUCATION, OUTREACH AND ADVOCACY.

THE LIVING BUILDING CHALLENGE WEBSITE

living-future.org/lbc

The online resource for project teams and others, it provides the Living Building Challenge standard document and the resources that support the certification process—including fee schedules for certification, detailed case studies of certified projects, and education resources. Detailed project team resources are available to registered project teams.

INTERNATIONAL LIVING FUTURE INSTITUTE MEMBERSHIP

living-future.org/membership

Access to Petal Handbooks is available to anyone with an International Living Future Institute membership. A current fee schedule is published on the Institute's website. Once logged in, members are directed to a unique homepage with links to update account details and access to the project registration form. The Dialogue is available only to registered project teams.

REGISTER A PROJECT

Registration is the first step toward Living Building Challenge certification and is accessible to ILFI members. Registration fees can be found on the Living Building Challenge website. The registration form contains prompts for basic information about the project, primary contact, owner and team. Most of the information provided at the time of registration can be updated, if necessary, by logging in to your membership page.

Registered projects can benefit from many Institute resources, such as the opportunity to submit program clarifications and exception verifications through the online Dialogue. They are also eligible to be added to the project team group account, the project team calls with the Living Building Challenge staff, and biannual in-person meetings. In addition, the Institute may contact project teams to showcase their work-in-progress through media outlets or in-house publications.

continued >>

CERTIFICATION OPTIONS:

Living Certification

Projects obtain Living Certification by attaining all requirements assigned to a Typology.

Petal Certification

Project teams may pursue Petal Certification by satisfying the requirements of three or more Petals (at least one of the following must be included: Water, Energy or Materials),

Net Zero Energy Building Certification

The Net Zero Energy Building Certification program requires four of the Living Building Challenge Imperatives to be achieved: 01, Limits to Growth, 06, Net Positive Energy (reduced to one hundred percent), 19, Beauty + Spirit, and 20, Inspiration + Education.

The requirement for Imperative 06, Net Positive Energy is reduced to one hundred percent, one hundred and five percent is required for Petal and Living Building Certification only.

All projects require twelve months of occupancy data before they can submit for certification. The exception is for a Petal Recognition project that is pursuing the Materials Petal and not the Water or Energy Petals.

Two-Part Certification is available for projects that wish to have a preliminary ruling issued on the Imperatives that are not reliant on occupancy data for certification. The preliminary audit may take place any time after construction is complete.

The following table identifies Imperatives eligible for preliminary audit and those requiring audit after the twelve-month occupancy period.

IMPERATIVE	PRELIMINARY AUDIT	FINAL AUDIT
01: Limits to Growth	X	
02: Urban Agriculture		X
03: Habitat Exchange	X	
04: Human Powered Living	X	
05: Net Positive Water		X
06: Net Positive Energy		X
07: Civilized Environment	X	
08: Healthy Interior Environment		X
09: Biophilic Environment	X	
10: Red List	X	
11: Embodied Carbon Footprint	X	
12: Responsible Industry	X	
13: Living Economy Sourcing	X	
14: Net Positive Waste		X
15: Human Scale + Humane Places	X	
16: Universal Access to Nature and Place	X	
17: Equitable Investment		X
18: Just Organizations	X	
19: Beauty + Spirit		X
20: Inspiration + Education	X	

continued >>

June Key Delta Community Center, Portland, OR
Photo: International Living Future Institute

The preliminary audit ruling does not constitute certification. Two-Part Certification is only possible when three complete petals have been achieved, one of which is either Water, Energy, or Materials. The audit is intended simply to give the team assurance that the Imperatives reviewed are in compliance with the requirements and anticipated for certification. The ruling on each Imperative will be carried forward to the final audit; however, if teams have completed work on the project that involved the use of new materials, an additional materials tracking sheet should be submitted outlining the materials used, listing compliance with Imperatives 11, 13 and 14.

FINAL AUDIT

For most projects, the same auditor will perform both reviews, although this cannot be guaranteed. The final review will involve a ruling by the auditor for certification.

Petal Handbooks

The Petal Handbooks are intended to serve as a resource for project teams pursuing the Living Building Challenge. Because the Living Building Challenge program is continuously informed by the work that project teams are doing on the ground, the handbooks have been developed to clarify and consolidate the rules at a set point in time to provide a unified reference for project teams. They are periodically updated to include all current Dialogue posts. While the Petal Handbooks are an excellent reference tool, they should be used in conjunction with the Dialogue to ensure that the most up-to-date rulings are understood.

Dialogue posts listed in the handbook will be archived but will remain searchable online.

The Dialogue

The Dialogue is an online host for the transparent exchange of ideas between project teams and the Institute—it is the official venue to request feedback on proposed strategies for meeting the requirements of the Living Building Challenge. The Dialogue allows for current unknowns to be discovered and shared in real time as teams proceed with their projects and research. It provides teams with the flexibility to get information most relevant to their work, such as in-depth commentaries, compliance paths, clarifications and temporary exceptions.

Organized by the twenty Imperatives and filterable based on specific content, the activity in the Dialogue not only serves as a platform for distributing strategies for success, it also yields modifications to future releases of the Standard itself. In this way, the Dialogue captures the ongoing evolution of the Living Building Challenge and gives credit to the hundreds, if not thousands, of individuals who contribute to the process. Dialogue content is available to registered projects only.

Submitting for Certification:

When a project team is ready to submit their project for Certification, they should contact the Institute at certification@livingbuildingchallenge.org.

continued >>

University of Wollongong, New South Wales, Australia
Courtesy: The University of Wollongong

TECHNICAL ASSISTANCE

Because the Living Building Challenge defines priorities on both a technical level and as a set of core values, it requires an approach to design, construction and operation that is fundamentally different than the current conventional structure. The Institute wants every undertaking to be successful on multiple levels. It supports a project team's transformative process of adopting the principles of the Challenge by offering optional services that shift the mindset and provide practical knowledge.

In addition to the specific services noted below, the Institute can also fashion customized options to match a project's needs during the design phases. The project team administrator may inquire about or schedule technical assistance by emailing certification@livingbuildingchallenge.org.

In-house Workshops

The Institute offers optional, customized training as a service for organizations and project teams to ensure that everyone has a shared fundamental understanding of the Living Building Challenge or particular Petal area. Whether there is a specific area of interest or a desire for a private presentation of an established curriculum, the Institute can deliver customized educational sessions. The most common workshop requested is a full-day introduction to Living Building Challenge that also includes discussion of contextual information such as development patterns and density, and regulatory, financial, behavioral and technological barriers and incentives. More in-depth, Petal-specific workshops that focus on Water, Energy and Materials are also available.

Charrette Facilitation

To steer teams toward innovative yet feasible solutions for their Living Building Challenge projects, the Institute offers an optional service to lead the kick-off meeting, or "charrette," and help define fundamental, strategic goals. A charrette should take place at the beginning of a project, when the potential to explore is at its fullest. The one-day meeting format focuses on fostering an interactive dialogue that allows participants to consider each area of impact. The two- or three-day format allows time for a deeper examination of promising ideas. The Institute designs the agenda, facilitates the session and provides a follow-up summary document.

Design Development Guidance Review

This optional service is intended to improve a project's potential to comply with the Living Building Challenge requirements at multiple points in the design process where adjustments are still possible. The Institute performs a remote review with the team to learn how the project accounts for each Imperative of the Living Building Challenge. Following a review of the project documents, the Institute will issue a report outlining our guidance for the team to improve their ability to succeed. It is possible to receive feedback on the Imperatives within a single Petal, select Petals, or all seven Petals of the Living Building Challenge.

continued >>

EDUCATION

The Institute is dedicated to transforming theory and practice in all sectors of the building industry, and offers several ways to broaden one's knowledge of deep-green building principles and practices, including the following:

Public Workshops + Webinars

The Institute offers in-person and online workshops taught by expert faculty about the Living Building Challenge and related topics. Workshops are continually developed throughout the year and are announced online and on the website. The Institute welcomes suggestions for future workshop content. Contact Institute staff to discuss options for hosting a workshop locally by emailing education@livingbuildingchallenge.org.

Living Future unConference

The Institute's three-day unConference is the flagship annual event for leading minds in the green building movement seeking solutions to the most daunting global issues of our time. Out-of-the-ordinary learning and networking formats deliver innovative design strategies, cutting-edge technical information, and much-needed inspiration to achieve progress toward a truly living future. Education sessions encourage a hopeful approach to the planet's economic, ecological and social challenges, and offer solutions for sites, infrastructure, buildings and neighborhoods.

Living Future offers project teams the opportunity to interact with other teams with similar project types, climates, or regulatory challenges. Each Living Future hosts a project team forum and several face-to-face gatherings.

continued >>

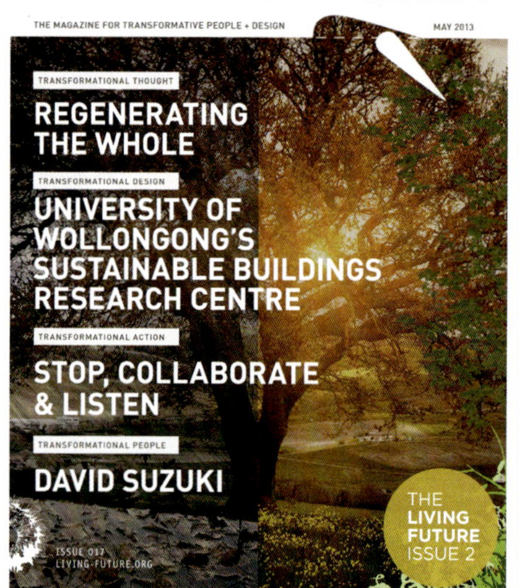

The Greenest Building: Quantifying the Environmental Value of Building Reuse

A REPORT BY:

WITH SUPPORT FROM:

IN PARTNERSHIP WITH:

Regulatory Pathways to Net Zero Water
Guidance for Innovative Water Projects in Seattle

Phase II Summary Report

February 2011

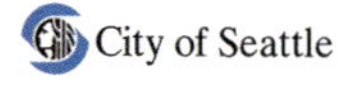

Prepared by:

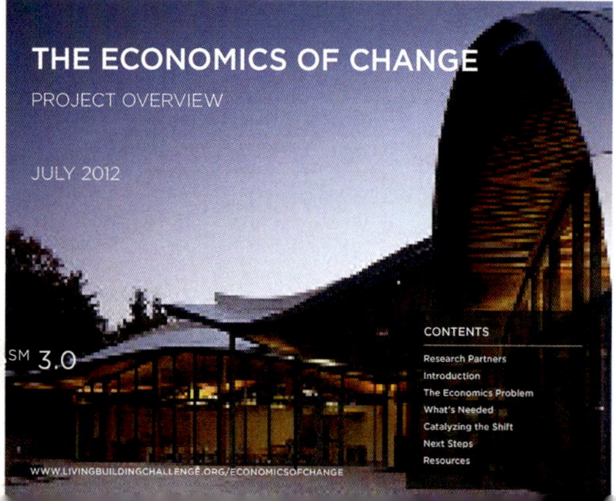

Trim Tab

Trim Tab is the Institute's quarterly digital magazine. Each issue features provocative articles, interviews and news on the issues, designs and people that are truly transforming the built environment. Subscriptions are free, and a complete archive of past issues is available on the Institute's website: **living-future.org/trimtab.**

RESEARCH

Despite the rigor of the Living Building Challenge, project teams are proving that the strict requirements of the program are very solvable. However, both perceived and real limitations to success still exist that are technical, regulatory, behavioral or financial—or a combination of these influencing factors. In collaboration with partners in the design and construction field, local and state governments, and other forward-thinking nonprofits, the Institute is spearheading efforts to carry out cutting-edge research and create practical tools. The latest published reports are posted on the Institute's website: **living-future.org/research.**

continued >>

AMBASSADOR NETWORK—SPREADING THE WORD ABOUT LIVING BUILDING CHALLENGE

The Ambassador Network is a global initiative to encourage the rapid and widespread adoption of restorative principles guided by the Living Building Challenge. Individuals from all walks of life are encouraged to sign up for our ambassador training program and to help us spread the word about a Living Future. The power of the network allows best practices and ideas to be shared globally, harnessing the best of social media and communication tools for rapid interchange. The network has been designed to support the continued flow of ideas and solutions among participants and the Institute. It presents numerous options for engagement, and
the Institute has created a wealth of related training materials and resources. More information about the Ambassador Network and the online applications are available on the Institute's website: **living-future.org/ambassador**

Ambassador Presenters of "An Introduction to the Living Building Challenge": Professionals who wish to shift the focus of green building conversations are trained through the Ambassador Network to deliver one-hour, informal introductory presentations to peers, local organizations, institutions, companies and community groups.

The presentations are delivered by volunteers, with the purpose of raising awareness around the Living Building Challenge. Ambassadors help build local capacity for the formation of a Living Building Challenge Collaborative.

Living Building Challenge Collaboratives: In communities all over the world, the principles of the Living Building Challenge are being shared and disseminated by our growing network of Collaboratives. These community-based groups meet in-person regularly to
share knowledge and create the local conditions that support development of Living Buildings, Sites and Communities. Each Collaborative is overseen by at least one trained volunteer facilitator, who is responsible for cultivating a welcoming environment for grassroots involvement and outreach. Each Living Building Challenge Collaborative has an online presence on Facebook and on the Living Building Challenge website. If there is no Collaborative in your area yet, we hope you will be inspired to form one.

OTHER WAYS TO GET INVOLVED

Continued advancement of the Living Building Challenge will require many minds and great ideas. The Institute has established a presence on an array of online communication forums that make it possible to aggregate impressions, suggestions and insights—please reach out to us today to get involved and contribute to a Living Future!

 /livingbuildingchallenge and /livingfutureinstitute

 @livingbuilding and @Living_Future

Okanagan College, Kelowna, BC
Courtesy: CEI Architecture

GLOSSARY

Adaptable Reuse
The process of reusing a site or building for a purpose other than the original purpose for which it was built or designed.

Adjacent properties
Properties or developments that share a property line with the project.

Black-water
Discharged water containing solid and liquid human wastes from toilets and urinals.

Brownfield
With certain legal exclusions and additions, the term "brownfield site" means real property, the expansion, redevelopment, or reuse of which may be complicated by the presence or potential presence of a hazardous substance, pollutant, or contaminant. Brownfields are designated as such by the EPA, or equivalent State, County or other jurisdictional body.

Chemical Abstracts Service (CAS) Number
A unique numerical identifier for nearly every known chemical, compound or organic substance.

Closed Loop Water Systems
Systems in which all water used on a project is captured, treated, used/reused and/or released within the boundaries of the project area.

Combustion
Any burning or combustion of fossil fuels or wood products.

Consumables
Non-durable goods that are likely to be used up or depleted quickly. Examples include office supplies, packaging and containers, paper and paper products, batteries and cleaning products.

Deconstruction
The systematic removal of materials from a building or project for the purposes of salvage, reuse and/or recycling.

Diverted Waste
All items removed from the project, including materials that are recycled, reused, salvaged or composted.

Dune
A sand hill or sand ridge formed by the wind, usually in desert regions or near lakes and oceans.

Durables
Goods that have utility over time rather than being depleted quickly through use. Examples include appliances, electronic equipment, mobile phones and furniture.

Energy Needs
All electricity, heating and cooling requirements of either grid-tied or off the grid systems, excluding back-up generators.

Floor Area Ratio (FAR)
FAR = Gross Building Area / Total Project Area

Forest Stewardship Council (FSC)
An independent, non-profit, membership-led organization that protects forests for future generations and sets standards under which forests and companies are certified. Membership consists of three equally weighted chambers—environmental, economic, and social—to ensure the balance and the highest level of integrity.

Greyfield
A previously developed property that is not contaminated to the level of a brownfield.

Greenfield
Land that was not previously developed or polluted.

Grey-water
Water discharged from sinks, showers, laundry, drinking fountains, etc., but not including water discharged from toilets and urinals.

Halogenated Flame Retardants (HFRs)
HFRs include PBDE, TBBPA, HBCD, Deca-BDE, TCPP, TCEP, Dechlorane Plus and other retardants with bromine or chlorine.

Land Trust
A nonprofit organization that, as all or part of its mission, actively works to conserve land by undertaking or assisting in land or conservation easement acquisition, or by its stewardship of such land or easements.

Landscape Succession
The gradual evolution of vegetation towards a more complex and ecologically appropriate state.

Manufacturer Location
The final point of fabrication or manufacture of an assembly or building material.

University of Wollongong, New South Wales, Australia
Courtesy: The University of Wollongong

Materials Construction Budget
All the material costs delivered to the site, excluding labor, soft costs and land.

Native Prairies
Diverse ecosystems dominated by grasses and other flowering plants called forbs; for the Challenge, Native Prairies can be either 'landscape remnants' or 'landscape restorations.'

Naturalized Plants
Plants that were introduced but are established as if native. Invasive plants that endanger native plants or ecosystems are not considered naturalized for the purposes of the Challenge.

Old-Growth Forest
Natural forests that have developed over a long period of time, generally at least 120 years, without experiencing severe, stand-replacing disturbance such as a fire, windstorm, or logging. Ecosystems distinguished by old trees and related structural attributes that may include tree size, accumulations of large dead woody material, number of canopy layers, species composition, and ecosystem function.

On-site Landscape
The planted area not used to comply with the requirements of Imperative 02: Urban Agriculture. The strategies implemented for each Imperative are not required to be mutually exclusive or physically separated.

Potable Water
Water that is fit for human consumption.

Previously Developed
A site with existing or historic structures or on-site infrastructure, or a site that has experienced disturbance related to building activity, including monoculture agriculture. Roads built for natural resource extraction (e.g., logging roads or mining areas) do NOT qualify a site as previously developed.

Primary Dune
A continuous or nearly continuous mound or ridge of sand with relatively steep seaward and landward slopes immediately landward and adjacent to the beach and subject to erosion and overtopping from high tides and waves during major coastal storms. The inland limit of the primary frontal dune occurs at the point where there is a distinct change from a relatively steep slope to a relatively mild slope.

Prime Farmland
Land that has been used for irrigated agricultural production at some time during the four years prior to the relevant Important Farmland Map date and where the soil meets the physical and chemical criteria for Prime Farmland or Farmland of Statewide Importance as determined by the USDA Natural Resources Conservation Service (NRCS).

Project Area
The entire scope of the project and all areas disturbed by the project work including areas of construction, staging and conveyance, which is typically, but not necessarily, all land within the property line. Project Area must be consistent across all Imperatives.

Project Water Discharge
All water leaving the building including stormwater, grey-water and black-water.

Renewable Energy
Energy generated through passive solar, photovoltaics, solar thermal, wind turbines, water-powered microturbines, direct geothermal or fuel cells powered by hydrogen generated from renewably powered electrolysis. Nuclear energy is not an acceptable option.

Salvaged Materials
Used building materials that can be re-purposed wholly in their current form or with slight refurbishment or alterations.

Stormwater
Precipitation that falls on the ground surfaces of a property.

Systems Furniture
A modular furniture system that might include work surfaces, cabinetry, file systems, flexible partitions and desk chairs used to create or furnish a series of offices workspaces.

Wetland
Those areas that are inundated or saturated by surface or groundwater at a frequency and duration sufficient to support, and that under normal circumstances do support, a prevalence of vegetation typically adapted for life in saturated soil conditions. Wetlands generally include swamps, marshes, bogs and similar areas.

Bertschi School, Seattle, WA
Living Certification - Living Building Challenge 2.0
Courtesy: space2place

Okanagan College, Kelowna, BC
Courtesy: CEI Architecture

VanDusen Botanical Garden
Visitor Center, Vancouver, BC
Photo: Nic Lehoux / Courtesy: Perkins+Will

David and Lucile Packard Foundation, Los Altos, CA
Net Zero Energy Building Certification
Photo: Jeremy Bitterman

NOTES

Phipps Conservatory and Botanical Gardens, Pittsburgh, PA
Net Zero Energy Building Certification
Photo: Denmarsh Photography, Inc.